EXECUTIVE DEVELOPMENT

EXECUTIVE DEVELOPMENT

Preparing for the 21st Century

HARPER W. MOULTON
ARTHUR A. FICKEL

New York   Oxford
OXFORD UNIVERSITY PRESS
1993

Oxford University Press

Oxford   New York   Toronto
Delhi   Bombay   Calcutta   Madras   Karachi
Kuala Lumpur   Singapore   Hong Kong   Tokyo
Nairobi   Dar es Salaam   Cape Town
Melbourne   Auckland   Madrid

and associated companies in
Berlin   Ibadan

Copyright © 1993 by Oxford University Press, Inc.

Published by Oxford University Press, Inc.,
200 Madison Avenue, New York, New York 10016

Oxford is a registered trademark of Oxford University Press

Library of Congress Cataloging-in-Publication Data
Moulton, Harper W.
Executive development : preparing for the 21st century / Harper W.
Moulton, Arthur A. Fickel.
p.   cm.
Includes bibliographical references and index.
ISBN 0-19-507465-3
1. Executives—Training of.   2. Assessment centers (Personnel
management procedure)   3. Industrial management—Study and teaching
(Higher)   I. Fickel, Arthur A.   II. Title.
HD38.2.M68   1993
658.4'07124—dc20   91-47136

987654321

Printed in the United States of America
on acid-free paper

To our wives,
Shirley Moulton and Sally Fickel,
whose support and patience have seen us through.

PREFACE

This book addresses the concept and issues of executive development and education, past, present, and future. The focus is on what the future requires in terms of executive competencies and on the implications of those requirements for the processes of life-long executive learning and development. To accomplish that focus,

- It summarizes the changes in the business environment both occurring today and anticipated in the future,
- It surveys developmental influences on executives and the causes for their success or failure, and
- It draws conclusions as to what those influences mean for future executive needs and how those needs can be met.

Executive development is a process as well as a product. It takes place primarily on the job and under the weight of on-going managerial responsibilities. Executive education plays a critical role in that development process because it makes experience more meaningful.

The processes of learning and development have been in place since the origins of humankind, and they continue to be central to the survival of institutions. During these final years of the twentieth century, change is occurring at an accelerated pace. That pace of change imposes new imperatives for corporations and executives worldwide to address the increasingly important issue of ensuring the competence of the critical executive resource.

We see three participants addressing this need for continuing long-term development of executives: first, the corporation, by reflecting in its policies commitment to executive development; second, the executive himself or herself, through dedication to life-long learning both on and off the job; and third, the directors of executive programs, in offering relevant and meaningful educational experiences to enhance the development process for program participants.

Executive programs encompass a relatively new branch of education that has emerged in the years since World War II. These mid-career learning experiences are offered both by universities and other institutions, and increasingly by large companies and government on an in-house basis. Practically all large and medium-size corporations partici-pate in public university executive programs, and projections indicate their steadily increasing use into the next century. Similarly, we see a marked growth of internal corporate programs focusing on company-specific as well as broader business issues.

Executive programs deal with the knowledge, skills, and perception required of business leadership. The typical general management pro-gram addresses the economic, social, cultural, technological, and politi-cal environments of business, as well as the ethical concerns of manage-ment. It also covers management process issues such as organizational behavior, planning and control, and key functional areas such as fi-nance, marketing, and operations.

Faculties for these programs consist of excellent teachers, inspiring motivators, and caring professionals drawn from academia, business, and government and consulting organizations. Executive attendees are experienced and successful people from middle, upper-middle and se-nior management positions in their organizations. Executive programs are typically residential, full-time, intensive, and expensive. Are they worth it? What makes some more effective and others less so? Under what circumstances? These and other questions relating to the value of executive education are addressed in this book.

From our experiences we have observed that most companies work hard to ensure the continuing growth and excellence of their executive resources; that most managers want to grow in competence; and that most managers are willing to accept a little executive education on the side as long as it does not interfere with their work and looks good on the record. Yet we have noted that most managers give little thought to what their own career patterns and aspirations call for in terms of new competencies, to the processes of learning and development, and to the role of life-long education in those processes.

In this book, we want to share with you our observations and insights regarding the processes of executive development and education. Our hope is that they stimulate you to think about how they might apply to you personally and to your organization.

The need for managerial and leadership competencies keeps growing and changing. Your own needs for development reflect the requirements of your increasing levels of maturity and responsibility. The world keeps shifting under your feet. We say that experience is the best teacher; but that is true only if you truly learn from it, and not only from your own but especially from the experience of others. The executive classroom can provide the opportunity for both kinds of experiential learning. How do you learn and what should you learn? How can an organization use the educational mode to foster some new corporate policy or thrust, not by indoctrination but by problem-oriented creativity in the classroom itself?

The worth of any educational experience is in part a function of how well it is designed and presented. But it is equally a function of how ready its beneficiaries are to receive it, which, in turn, depends on how well they understand the kinds of issues we discuss in this book. Furthermore, any manager will be better able to take charge of his or her own development and career by thinking about those issues. Therein lie the reasons for our writing this book.

Saunderstown, R.I.   H. W. M.
Fort Belvoir, Va.   A. A. F.

ACKNOWLEDGMENTS

We can think back to a score of outstanding professors and other resource people in our executive programs at General Electric, Crotonville, and IBM with whom we spent so much time both in and outside the classroom, as well as to so many great students with whom we worked. We also think of some fifty books and journals that we selected to stock the mini-libraries in our students' dormitory rooms. Most of all, we think of our thirty years of experience in a variety of operational and staff situations, and of the lessons we learned about the behavior of individuals, organizations, and forces in society.

We wish to acknowledge the following sources of data, ideas, and insights that have contributed significantly to the content of this work.

Institutions and Organizations

The Institute for the Future (Menlo Park, California) for a greater understanding of future trends

Drexel University for their program in "Futures"

National Industrial Conference Board for their insights in the field of executive education

The U.S. Naval Institute for a greater understanding of the role of executive education in U.S. corporations

Surveys and Studies

Retail Recruiters International Survey, highlighting career problems for middle managers

McKinsey survey of trends in international sales by U.S. companies

Heidrick & Struggles' study of executive mobility

Publications

Numerous articles from the following publications have contributed significantly to the content of this book: Fortune, Harvard Business Review, Time, U.S. News and World Report, Business Week, Nations Business, Forbes, and Modern Maturity. In addition, we consulted *The New York Times* and *The Providence Journal–Bulletin* newspapers on a regular basis.

Individuals

Peter Drucker, one of the best known management authorities, from numerous articles, speeches, lectures and personal conversations

Harry Levinson, internationally known psychologist and management consultant, from personal contacts as well as numerous articles and lectures, on executive stress

Eugene Jennings, leading management consultant, from years of personal contact, speeches, and articles, on the developmental aspects of mobility

James Giffin, vice-president, Aetna Life, on the federal deficit

Jacob Bronowski, Center for Creative Leadership, on experiential learning

Malcolm Knowles, well-known author and expert on the principles of adult learning

John Gardner, well-known author, on the issues of personal development and lifelong learning

Abraham Zaleznik, noted behavioral scientist, for his views on the stresses and responsibilities of leadership

Walter Wriston, past chief executive officer of Citicorp, on the fallacies of protectionism

Akio Morita, Sony chairman, on the economic system and federal deficits

Mitsubshi chairman, on managerial performance and development

In addition, we wish to thank Alan White, senior associate dean, MIT; Ron Bendersky, associate director of executive education, University of

Michigan; Al Vicere, assistant dean for executive education, Penn State University; Bob Fair, professor, University of Virginia; and particularly Ray Watson, past director of executive education at Penn State University and Duke University, and current board member of the University Consortium for Executive Education, for their support and valuable counsel in the preparation of this book. Last, but not least, our gratitude goes to Peter Crolius, publisher, Dutch Island Press, for his valuable advice and to Peter Vogt, our word processing expert, without whose hours of work and dedication this book would never have been completed.

CONTENTS

Part II Lifelong Learning

Part I

Executive Competency

1

The Business Environment in the Twenty-first Century

What will the world be like in the year 2000 and beyond? Will it be a nuclear wasteland? Barring that scenario, what will the business environment be in the next century? What will the major components of that environment be? The global economy, the worldwide political scene, social developments, and the level and impact of technology immediately come to mind.

Why do we need to study the future? We need to because a perceptive awareness of the changing environment is vital in seeking direction for managerial development.

How We Study the Future

Since the 1970s, studies of the future have been very much in vogue. Such organizations as the Rand Corporation, the Hudson Institute, and the Futures Group, among others, have been extensively involved in studies conducted for the military, business, and other organizations. The purpose of such studies is to help those institutions with their strategic planning, an increasingly important activity within most major corporations.

There are two fundamental questions involved in any consideration of the future: What are the processes to be used in thinking about the future, and what are the basic forces that will be driving the global system in the next century?

One answer to the first question is the use of forecasting, that is, using current information to project future events. However, as pointed out by Pierre Wack (1985), former head of Royal Dutch Shell's Strategic Environment Unit, forecasts are dangerous because sometimes they work, and sometimes they don't (as in the case of the OPEC-induced oil crisis of 1979). Sooner or later forecasts will fail, and they fail when you need them most, because they do not anticipate major shifts in the business environment that make corporate strategies obsolete. According to Wack, it is impossible to forecast the future unless it is predetermined. Therein lies the challenge: the process of futures research is a function of identifying the predetermined things and the unknown things, which can be done through the creation of scenarios.

State-of-the-art activity in futures research lies in the creation of scenarios, which are stories about the future . . . what might happen, not what will happen. Scenarios involve two fundamental dimensions:

1. Predetermined things: events and developments that result from actual circumstances and that therefore cannot be changed in any significant way.
2. Uncertain things: circumstances that are not known and that are, therefore, unpredictable.

An example from nature should suffice to illustrate what we mean by "predetermined." Years ago in India, extraordinarily heavy monsoon rains occurred in the Himalayas, in the upper part of the Ganges Basin. It could be anticipated with certainty that within two days serious flooding would occur in Rishikesh at the foothills of the Himalayas, seven to nine days later in Allahabad, and twelve days later in Benares. This sort of prediction is not crystal ball gazing, but simply describes the future implications of something that has already happened. Therefore the first approach to scenario planning is to identify such predetermined elements, which can be seen as an answer to the second question about considering the future.

Predetermined Things—A Summary*

In times of fast-paced change, existing conditions can, in some respects, become quickly outdated. Therefore, it is not the topical events or conditions themselves but rather the on-going significant trends or underlying forces they illustrate that we hope the reader will seek to

*A more extensive discussion appears in Appendix A.

identify in the following summary. Their significance for the future executive lies not in the static descriptions of predetermined things but in the dynamic forces behind them. It is in that sense that we use the cryptic term "predetermined things." We can note them, but the reader or executive must evaluate their relative significance as he or she sees them.

The scenario building process has as its first step a listing of those events and developments that are already in place, and that are likely to continue into the future. The following material is a summary of major changes already in the pipeline that must be considered in any thinking about the future. When integrated into the three scenarios presented later in this chapter, these considerations should provide a reasonable basis for developing a projection of requisite executive competencies for the next century.

Social/Cultural

We are looking at a 6.35 billion world population by the year 2000, reaching 10 billion by 2030. Within this population, there is dramatic growth in aging segment, particularly in developed nations: 24 percent in the United States, 30 percent in Japan, and 36 percent in West Germany will reach retirement age by the year 2000. Also in that year, the U.S. labor force will be 25 percent black or Hispanic with a growing percentage of minority group and women working. These figures mean a dramatic change in the management process and the makeup of the managerial group will be needed.

Environmental

Energy and pollution will continue to be a serious problem for political and business leadership in the future. Alternative energy sources must be developed that are nonpolluting. International business must play a significant role in the reduction of pollutants and in finding more efficient methods of energy utilization in production processes.

Political

The year 1989 will be remembered as a major turning point in the world political environment for the decline of communism, the rise of democracy, and the advent of market economies in eastern Europe. Even more dramatic have been the wrenching changes in the Soviet Union in 1991, resulting in the collapse of the Soviet system after seventy years of communist rule. Severe economic conditions and political problems will be inevitable in the short run. In the longer term, there will be a need for increased investment in that country as it slowly moves to a free market economy. A dividend should come from a reduction of conventional and nuclear forces in both the new USSR and the United States, resulting in sharp decreases in defense budgets. These developments along with possible reductions in trade barriers, in U.S. trade and federal deficits, and in regional instabilities, should lead to promise for future economic growth.

Business Activity

The downsizing and restructuring of U.S. businesses has been dramatic and effective in increasing global competitiveness for America. Likewise, the ongoing process of mergers, consolidations, and takeovers should, on balance, result in tighter management and more efficient use of resources. Business ownership is becoming thoroughly international, which should underscore the development of a truly global economy that is becoming increasingly service oriented and entrepreneurial in nature. The implications for management are crystal clear: global strategic thinking, international orientation, lean and efficient operations, and long-term commitment to growth and profitability will be requirements for survival and success.

Technology

Probably one of the most significant developments in technology is in the field of telecommunications, information systems, and computers. Technological literacy is already impacting business functions and how managers make decisions, and will be essential for executives and managers at the end of this century. Automation and robotics will continue to play a significant role in strengthening both the production and service sectors of the world economy.

This summary of predetermined things includes social/cultural, environmental, political, business, and technological developments that are currently in place and that will continue to be influential in determining the future world economy. Therefore, economic conditions in themselves are not included in this category of predetermined things. They constitute the uncertain things dealt with through the creation of the three long-term economic growth scenarios described in Chapter 2.

Good managers respond well to change because they understand it. Good leaders understand change because they create it. The best of managers and leaders anticipate change, not only in the world about them but also within themselves. Nobody can keep up with all the details or define all the consequences. Even between our writing and your reading of this book much has changed. But you can keep abreast of major trends, absorb new ideas and values, and continually prepare for a constantly changing future through personal vigilance and study that will enable you (and therefore your organization) to survive and prosper.

Uncertain Things

The preceding summary of predetermined things includes changes that will be significant for executives over at least the next several decades. Those changes are fairly predictable because they are already firmly established as the result of fundamental and observable trends and therefore are not likely to be reversed.

But another category of change, the one that we call "uncertain things," also significantly affects executives, yet is considerably more difficult to project into the future with any degree of confidence. This kind of change gives rise to the well-known observation that reasonable men working from the same facts can reasonably reach quite different conclusions.

Uncertainty about the future lies mostly in an area of human activity having roots that reach out in many directions of change: social, political, physical resources, technological, psychological, and so on. In a democratic society that activity manifests itself primarily in what we call the economy, or economic activity. The economy is of lively and crucial concern to everyone—national leadership, institutional leaders, and the man in the street, all alike. Yet it is the essence of uncertainty.

There is no better way to illustrate this phenomenon of uncertainty than to recall how professional and respected economists—the people who presumably have a better grasp of the subject than do the rest of us—differ in their analyses and theories of the future based on their evaluation of the past. At least four schools of economic thought emerge. One school says that we shall gradually return to the "normalcy" of the 1960s. A second says that nothing has changed or will change; cycles of recession and inflation have always existed and will continue. A third school, taking a creative approach in lieu of prediction, says that we need a "restructuring" through new industries in the private sector; a reworking of the public sector to deal with deficits while repairing the infrastructure; and new arrangements with lesser developed nations. A fourth school of thought throws up its collective hands in despair and declares that the economy is out of control. They say we've come forth with fixes too little and too late, that there are enormous problems in the international financial system and liquidity, and that any improvement will take a long time.

Economic Scenarios

Who is right? Recognizing the various opinions of economists, futurists do not try to answer that question categorically. Instead, they construct a series of landscapes of the future—scenarios—that, in their opinions, are all-inclusive, internally consistent, logically reasonable, and relevant to the purpose being served; that is, in this case, to help identify the kind of executive competencies that are likely to be required in the future. Thus the problem of anticipating competency requirements becomes one of making a choice among several scenarios based on intuitive judgement as to which one is more probable.

Following are three scenarios scaled along the dimension of relative attractiveness: most pessimistic to most optimistic. We are making the hopeful assumption that relative attractiveness correlates positively with relative probability. That is our American spirit at work.

Scenario 1

"Divided World Scenario," also known as the "World of Internal Contradictions," "Muddle-Through Scenario," or "Restrained-Growth Scenario."

Linked to conventional business cycle

Growth forces not liberated—growth low and sporadic

Traditional systems overexpand and decay

Alienation and social strains

Societies "close up" to avoid risk

Protectionism and government intervention

World remains divided

Most pessimistic scenario.

Scenario 2

"Renewed Growth Scenario," also known as "Next Wave Scenario."

Linked to "boom-bust" business cycle

Disturbed world gaining balance

Managerial challenge of managing extremes

Renewed growth and fuller resource utilization

High innovation, savings, and investment

Crisis-based management, with high levels of pain and stress

More optimistic scenario.

Scenario 3

"Solid Growth Scenario," also known as "One-World Scenario"

No "boom or bust" business cycle swings

Steady, manageable growth

Interdependence of nations

Decline of protectionism and rise of international trade

Increasing stability of international monetary system

Federal government budget cutting and deficit reduction

Rising employment, wealth, productivity, and efficiency

Increasing growth of service economy

Most optimistic scenario.

Other scenarios could be developed, but these three represent a spectrum of realistic possibilities, running from extreme pessimism to optimism. In some sense "you pays your money and takes your choice." In view of political, economic, and social conditions in the early 1990s, we are inclined to choose scenario 3 as the basis for our expectations concerning the future business environment and the subsequent requirements for executives.

2

The Twenty-first Century
Executive

The year 2000 holds a special fascination for Americans. Futures studies have gained prominence over the past several decades and continue
to be a subject of growing interest as, historically, there has always been
an upswing in interest at the end of each century.

Currently, about 1,000 courses on the future are being offered at
schools and universities in the United States, from Drexel University
offering a course on the sociology of the future, to the University of
Houston offering a master's degree in futures studies. On the corporate
side, General Motors; Sears, Roebuck, and Co.; and others have turned
to futurists to help develop competitive strategies. Such think tanks as
the Futures Group in Glastonbury, Connecticut; the Hudson Institute in
Indianapolis, Indiana; and the Institute for the Future in Menlo Park,
California, play a significant role in futures research. Books such as
Megatrends, Future Shock, and others continue to sell well. The World
Future Society, a nonprofit group devoted to clearinghouse operations in
futures forecasting, has over 25,000 members. In 1960, futurists were
seen as science fiction addicts and hobbyists, and as unnecessary. Today
they are taken seriously, as evidenced by former President Reagan having met with a group of futurists in the White House.

One of the reasons for this growing interest in the future has been the
rapid pace of social and technological change, which leads people to
want to understand those forces affecting their lives today and in the
coming decades. Twenty-first century challenges include intense competition, increasingly global markets, and rapid technological change.
How many executives really appreciate what it will take to succeed in
the future? It is likely that the underlying issue will be one of managerial
competence (knowledge, skills, and attitudes).

This chapter is intended to provide perspective on and identify those competencies that have been, and continue to be, requirements for effective managerial performance in today's world, and which will be of increasing importance in the future. The remaining chapters of this book provide an evaluation of how these competencies can be enhanced through meaningful on-the-job experience and formal executive development and education programs, and discussions about the internal and external programs themselves.

Trends Supporting Executive Competency Requirements

You should consider the following trends regarding changes in the business and managerial environment as shifts in emphasis only, and not as absolutes. Furthermore, we do not intend to indict the past but rather to complement it, because most of these shifts have been underway for some time. We foresee the changes of necessity accelerating.

1. The scope of business operations is changing. The tactical, short-term focus with its emphasis on quarterly profit and loss figures is gradually giving way to a strategic, long-term perspective. The focus is moving from a primary concern for national marketplaces to a realization that the future survival of business depends on an international, worldwide set of customers. More and more, we see a truly global economy that is interrelated and interdependent. Continued growth rests on increasing investment in research and development, new high quality products, tight control of costs, and innovative marketing approaches to meet the needs of an international marketplace.

2. The evolution of organizational forms is characterized by a shift from pyramidal, hierarchical structures to flat, decentralized entities. There is a growing emphasis on entrepreneurial freedom to manage proliferating independent units within the corporation. While functional dimensions of organizations continue to be important, the application of necessary functional expertise will increasingly decentralize to semi-autonomous business units. Overall integration will be achieved through information networks rather than by the imposition of authority and control from corporate headquarters.

3. The attitude of business executives toward the national and international corporate "stakeholders"—the employees, shareholders, suppliers, community, customers and government—is changing. The traditional passive, business-as-usual, indifferent attitudes, which reflect a singular lack of concern for these important constituencies, are giving way to a more responsible, proactive, and flexible approach to members of that external environment. In a recent study of executives of the largest oil companies, one of the major findings indicated that historical passivity and later stonewalling and confrontation with government officials, the news media and special interest groups was counterproductive. The study concluded that a more enlightened approach and sensitive concern for these constituencies was required.

4. The traditional decision-making process is shifting from a quantitative, numbers-driven orientation to a more qualitative mode, wherein the issues to be decided are based on value questions such as "what ought to be done," as well as "how can we do what we have been doing more efficiently"? Problem-solving may be giving way to "path finding," new ways to exploit opportunities. What is worth doing and how it can be done will be the decision focus in the future.

5. The requirements for management will shift from following to leading, from business-as-usual to innovation, from a reactive mode to a proactive stance toward business issues. The traditional win–lose philosophy will be replaced by a win–win attitude, in which negotiation and constructive compromise will be the name of the game. International corporate alliances and cooperative arrangements across borders will demand political skills, thus replacing the insular thinking that has dominated the past. Because of the growing importance of technology to business success, technological sophistication on the part of all managers will be a necessity.

Other trends are clear: smaller, leaner, decentralized entrepreneurial business units are proliferating, particularly in the high-tech and service sectors, requiring a proactive posture to shape the environment rather than passively reacting to it.

Finally, we can see that organizationally and geographically the United States is moving toward one of the most diffuse and decentralized economies in the free world. While no longer the dominant player, it will likely become a very strong and efficient trading partner in the global marketplace in the future. Such a change in international relationships will require new sets of executive competencies to manage in that world.

Personal Competency Requirements for Executives

The following competencies are an amalgam of skills, knowledge, and attitudes that we believe will be required in some combination by managers and executives in the next century. Skill without requisite knowledge and appropriate attitudes for its application is made useless. Rather than getting bogged down in a series of definitional questions as to what is a skill as a stand-alone concept, and how it can be developed, we suggest the term *competency,* encompassing skills, knowledge, and attitudinal components, as more useful.

Environmental: Global orientation, intercultural understanding, political sophistication, foreign language fluency

Leadership: Proactive leadership orientation, understanding values and ethics, innovative and creative ability, motivating others through a sense of mission

Managerial: Integrative ability, technological literacy, breadth plus depth of knowledge, flexible and adaptive behavior

Interpersonal: Negotiation and communications skills, emotional and physical fitness

Business implementation: Strategy formulation and policy development, functional sophistication, microeconomic literacy and appreciation

Scenarios and Competencies

As stated in Chapter 1, scenarios are stories about the future; they show what might happen, not what will happen. The predetermined things previously described must be taken into account in constructing scenarios, while the uncertain things are dealt with as alternative futures in the scenario format.

It would be useful to comment on the relevance of executive competencies required in the future for each of the three scenarios presented in Chapter 1. Each scenario suggests somewhat different trends in executive competency requirements.

Scenario 1

This "Divided World Scenario" suggests a return to short-term, nationally focused interests, and emphasizes an insular, reactive, and win–lose approach to business issues. Competencies called for would be political and economic toughness in dealing with increasingly nationalistic foreign and domestic governments because of growing concerns for protectionism on all sides. Leadership would call for an innovative and creative ability to deal effectively with a closed system to maximize efficiency of operations that are severely circumscribed. Managerial flexibility and adaptive behavior might prove important in dealing with a "status quo" business climate. Interpersonal and communications skills would come into play in maintaining morale within the organization and in attempting to influence government policy makers to see the increasing needs for growth of foreign trade.

Scenario 2

In a "Renewed Growth Scenario," national economies would be highly volatile, with sharp swings in the business cycle. Renewed growth would call for quick-acting business leadership in order to take advantage of new opportunities. Flexible and adaptive behavior by managers would be needed, as well as the interpersonal qualities of emotional and physical fitness to handle the stress of radical change.

Scenario 3

The most optimistic of the three scenarios, the "Solid Growth Scenario" indicates a more challenging and stable international business environment. It calls for the full range of the competencies previously listed, with special emphasis on the following: global and long-term orientation; ability to create and work with exotic financial structures; intercultural understanding and proactive leadership; the development of a philosophy of mission to energize human resources including foreign workforces; and foreign language fluency for executives dealing a growing, interdependent global economy.

Managerial Level and Competencies

The executive competencies listed in the previous section are most applicable to the more senior levels of management, although some of those competencies apply to all managerial work regardless of level. For example, first-level managers and managers of managers need leadership, decision-making, and interpersonal competencies at the lower organizational levels where the focus is on tactical operations and decisions. More senior functional and division managers have a much wider range of responsibility, requiring a broader, strategic perspective of their division and of the corporation. General managers and corporate officers spend a larger amount of their time with problems arising in the external environment, such as the community, national and international governments, stockholders, labor unions, and other constituencies. Senior managers are constantly concerned with long-range strategic planning, investments, succession planning, fiscal discipline, world markets, competitors, government relations and other environmental impacts on the on-going survivability and success of the organization.

Conclusions

The preceding list of competencies has two important characteristics. First, it is based on the previously developed trends and scenarios relating to the business and economic environment in the twenty-first century. Second, the list includes those competencies that can be enhanced or developed through experience on the job and through appropriate executive education programs.

The reader will note that we have not dealt with the all-important matter of the quality of personal behavior of managers, executives, and, especially, top leaders. Quality of behavior is largely evident as a result of personal traits, which should be distinguished from competencies. Competencies are skills, knowledge, and attitudes that can be taught and changed. On the other hand, traits are largely a function of individual personality and character. Examples of traits include integrity, initiative, aggressiveness, judgment, and decisiveness, among many others. Such traits are central to effective executive performance, and while not "trainable" as such, are instrumental to the effective application of competencies on a given job. Although that is a matter beyond the scope of this book, we mention it so that the processes of developing operating competencies may be considered in a proper context.

Regardless of which economic future presents itself and the organizational level of a given manager or executive, the requirements for continuing change and growth into the twenty-first century touch on every conceivable aspect of individual competency. Those competencies require a lifetime of effort, both on and off the job, to change and enhance them—the subject matter for the remaining chapters of this book.

3

Experience and Performance

> We have to understand that the world can only be grasped by
> action, not by contemplation. . . . The most powerful drive in the
> ascent of man is his pleasure in his own skill. He loves to do what
> he does well and, having done it well, he loves to do it better
>
> JACOB BRONOWSKI, "The Lessons of Experience"
> The Center for Creative Leadership

Self-Development Principle

In the final analysis, all development is self-development whether it
takes place on or off the job, at work or at home, in school or on
vacation. You have the ultimate responsibility for your continuing
growth as a person, a parent, a citizen, and as a professional manager or
executive on a defined career path. All experiences can and should
contribute to your life goals and career objectives, and your role in this
process is of primary importance.

Development is a growth process that takes place continually
throughout life from one stage to another. It is largely a function of
individual motivation, intelligence, receptivity, health, and attitudes
that can utilize all the external inputs you are exposed to throughout life
in a meaningful way. You should select, refine, and integrate experience
into yourself such that competence is enhanced and development is
continuous. It has been said that either you grow or die, which is
particularly true in a rapidly changing world. For example, the projected
half-life of an engineer's knowledge is four years, which suggests that
one half of this knowledge is obsolete in that time frame, and that a
continuing updating of that knowledge base must have top priority.

If we now move to the application of this principle of development as a continual growth process to the managerial environment, it is clear that the ongoing development of managers and executives is a critical concern to business leadership. The application of this principle takes place in a number of ways: job experience, interpersonal relationships, educational programs, and a number of off-the-job self-development activities. In today's corporate setting, as elsewhere, change is the name of the game. It can be viewed as a challenge or a threat, offering unparalleled opportunities for growth and development on one hand, and failure on the other. It is the individual manager's responsibility to respond positively to new demands for efficiency, down-sizing, restructuring, new technologies, global operations, and increasing foreign competition.

The life-long development of the whole person is the foundation for your development as a manager and a leader. Thus you must take charge and seek out the kind and quality of your life experiences, both on and off the job, that will advance rather than impede development. Because so much time is spent on the job, for most people working-life experiences are (or can be) the most developmental. We now examine the various dimensions of those experiences.

We know that on-the-job challenge is crucial to the development of managerial abilities. Meaningful experience is the key to this development process, not just experience in and of itself. It has been said that in certain instances a given manager has had ten years experience that could be translated into one year's experience repeated ten times— hardly developmental. We learn by doing our managerial work under the full weight of responsibility in an ongoing organizational environment; and we grow by being challenged to do new things in performing that work.

However, there are at least two aspects to experiential learning that must be considered. First, the nature of the experience must be evaluated to determine which experiences matter more than others and have potentially more impact on successful careers. (An analysis of key on-the-job experiences follows.) Second, the ability to learn from experience must be present. An open-minded attitude and a critical self-awareness are essential to reflect on the broader lessons of such experience. Many bright executives have been so concerned with the narrow aspects of performance that they have neglected or not seen the negative impact of their behavior on subordinates, peers, and superiors. Hence, the importance of feedback to give an accurate assessment of total performance should be recognized.

It should be stated at this point that the role of formal executive education is crucial to making this on-the-job learning more meaningful by providing renewal, perspective, and increased self-confidence as well as enhancement of the knowledge and skills so necessary for successful careers. We've heard it said that you can learn more about your job by being away from it. Executive education programs take the participants away from their jobs. The role of executive education is the subject of chapters 7 through 11.

Critical Assignments

As previously stated, the first aspect of experiential learning deals with the nature of the experience. The following analyses of the key on-the-job categories should lead to a greater understanding of the growth and developmental possibilities in each category.

Project and Task Force Assignments

Most managers and executives begin their careers in a technical or functional specialty, such as finance, engineering, or production, and success has been principally due to their technical excellence in those fields. Project and task force assignments place them in a situation where these old skills don't mean much, and others probably know even more than they do. Such assignments usually come on top of, and not in place of, an already challenging and demanding job. They are characterized by high visibility to senior management (with attendant high risk), and usually have a finite time frame of from six months to one year.

Such assignments usually fall into three categories: (1) installing new systems or trying out new ideas; (2) negotiating agreements with external parties; and (3) trouble-shooting a problem-filled situation. Regardless of the type of assignment, the manager or executive faces a dilemma: managing a situation where his expertise may be irrelevant, and dealing with others over which he has no control or authority. The developmental aspects are clear—coping with the unknown, dealing with groups of strangers, handling the pressure of highly visible success or failure.

The manager in this situation can do one of two things: try to take control by applying his own technical competence or by trying to master all of the technical detail of the new assignment (which would probably be counterproductive), or take the attitude that he might be able to make a contribution to a project without having mastery over every detail. This approach requires another kind of competence, that of understanding and listening to the point of view of others, and of working with them to arrive at a workable, meaningful solution to the problem at hand. Here, authority and control are not the issues; persuasion, negotiation, and patience are. Listening, asking questions, and seeking consensus are the keys to achieving a common goal.

The payoff can be significant: enhancement of management skills, an increased sense of self-confidence, and experience in dealing successfully with the unknown—important preparation for higher managerial responsibilities.

Line to Staff Assignments

Despite the on-going reduction of corporate staff positions in U.S. business enterprise, staff will continue to exist, contributing to such needs as strategic planning, financial control, and human resources coordination. A one or two year assignment to a corporate staff position can provide a critical developmental experience for a line manager whose previous career has been restricted to operations. The shift can be traumatic. The manager moves from a position of bottom-line numbers and line authority, to an assignment fraught with ambiguity and anxiety. In this situation, a manger is exposed to senior management on an almost daily basis, and is frequently required to make presentations to an executive group on subjects that were mostly unknown six months or a year prior to the assignment. The job must be learned, relationships must be understood, and priorities must be established, all within a context often without well-defined limits.

The benefits of such assignments are many: development of an understanding of corporate strategies and culture, access to policy and top management concerns, exposure to privileged information, and high visibility to key executives on a regular basis. Diplomacy and tact are fundamental to effective performance because getting the cooperation and information from business unit executives that is often necessary to meet staff objectives must be done without the clout of line authority. In addition, the staff assignment will provide exposure to external forces that impact the business, such as government officials, labor leaders, competitors, stockholders, Wall Street analysts, and others whose interests must be taken into account in the on-going strategic directions of the business.

Learning and development (and education, for that matter) are uncomfortable and even stressful experiences, and staff assignments are no exception. Many line managers have difficulty adjusting to staff work, but most find that the rewards from successful completion of a new and difficult corporate staff assignment will play a major role in continued career success in subsequent line executive positions. Perspective, maturity, and new levels of confidence can't help but contribute to greater competence in dealing with the challenges of the future.

Assistant-to Assignments

IBM and many other large corporations utilize the concept of assistant-to positions as a major developmental device for young, high-potential managers. These managers typically have had one or two line management assignments, and, as part of their career planning within the corporate succession tables, will be assigned as either an administrative or executive assistant to a major executive, usually at the corporate level. Such assignments are usually for one year, but could be somewhat longer depending on the next available line position at a higher managerial level.

A great deal of care should be taken in matching the young manager with the appropriate executive so that the experience is of maximum benefit to both parties. Senior executives vary in management style, particularly in terms of how they work with supporting staff and their peers. The most likely candidates for sponsoring an assistant operate in an open manner, delegating responsibility and sharing information. In such cases, the assistant can play a variety of highly developmental roles, from opening mail and checking the executive's in-basket, to attending meetings with her, taking trips with her, and participating in planning, problem-solving, and strategy discussions.

Thus, assistant-to positions can be educational and developmental if well planned. They can provide the opportunity for exposure to the top levels of the business, to a strategic perspective on corporate operations, to the management style of a key executive, and to the opportunity to participate in a meaningful way in the management process at senior levels. Almost every senior executive in the IBM corporation has held assistant-to positions at early career stages, which has enriched their experience base and helped to accelerate career progress.

It should be noted, however, that the case in favor of assistant-to positions as a developmental device contrasts with an equally strong opposing view held by some companies (notably, General Electric). They feel that assistant-to and deputy positions more often than not are harmful to the individual, are a source of confusion for the organization, and are an inefficient use of a critical resource. This view holds that such positions must be used (if at all) only in special and quite temporary circumstances. How effective assistant-to positions are as developmental devices is likely to be a function of the ethos of the organization and of its management philosophy.

International Assignments

As business activity becomes more global, the need for managers and executives with international experience becomes essential to corporate competitiveness and long-term positioning in the world marketplace. Most leading multinational corporations in the United States, Europe, and Far East have established policies of staffing overseas branches, subsidiaries, and regional or area headquarters organizations with an appropriate mix of executive talent from both the host country and the home country. While there is much to be said for utilizing foreign nationals to manage and operate overseas business units, it is vital for representatives from the parent company to have hands-on experience in a foreign country through planned rotational assignments lasting one, two, or more years in line, staff or consulting management positions in the major countries in which the company does business. It is equally important for line and staff executives from the overseas country to have the opportunity to work at corporate headquarters and/or major business units in the home country for experience and perspective. They thus gain exposure to the top management team, and a better understanding of the strategic directions of their company and the key role of foreign nationals in managing a complex international business.

The benefits of international experience are myriad. The values, attitudes, and operating methods of other cultures must be incorporated into a worldwide policy structure, leading to a truly international business philosophy. Such cultural assimilation can only be achieved through extended overseas experience and not by occasional one- or two-day flying trips. The learning opportunities for coping with strange environments, different values, and other language and management styles can be invaluable. Relationships with peers from other countries must be cultivated and mutual understandings need to be developed, as common corporate interests and policies will likely require interpretation and reasonable adaptation to local circumstances. Flexibility, adaptability and sensitivity, and respect for differences are basic requirements for success in such international assignments.

Selection for overseas posts must be carefully done. Not only must the candidate for such assignments be willing and available to spend several years away from his home country, he should also understand that this new experience will contribute to individual growth and maturity, as well as to future career success. The corporation will also benefit by having an executive team with the hands-on international experience and worldwide business perspective so necessary to effective global operations.

The family of an employee who is assigned overseas will play a key role in that individual's success or failure. The investment in moving a family overseas is not trivial. A two year assignment in South America, for example, will cost the company hundreds of thousands of dollars, apart from salary and benefits. If the wife and children of the assignee are unhappy, the chances for failure are high. In one major international company, the family is flown to the overseas location to check out housing, schools, health care, and other conditions before the final decision is made. Families of managers who have had a number of previous assignments and physical moves are better choices for overseas positions than those who have remained in one location.

If at all possible, foreign language skills should be required, or at least an intensive course in the language should be undertaken before posting the overseas assignment. While recognizing that the English language has become the international standard for global business, fluency in the host country's idiom provides an enormous advantage to effective functioning in a foreign environment. Even without fluency, a manager who tries to speak the other language can generate enormous goodwill with international colleagues.

Finally, a career manager should be in place to keep track of overseas performance, maintain contact with the assignee, and help assure an appropriate reentry position for the individual so that such an assignment is consistent with corporate staffing objectives and career planning. In view of the total investment by the company, the manager, and his family, such assignments should be carefully planned and monitored to assure the long-term payoff for both the individual and the corporation.

Developmental Line Assignments

While staff, project, assistant-to, and foreign assignments feature persuasion, analysis, ambiguity, and exposure to power, line management is where the action is—where "the rubber meets the road"—and the lessons of accountability and the exercise of power are manifest. According to the authors of a recent study by the Center for Creative Leadership (McCall and Lombardo, 1988), *The Lessons of Experience,* three major categories of line assignments offer the best opportunities for developing leadership: starting something from scratch, turn-around or "fix-it," and managing an operation of larger scope.

START-UP ASSIGNMENTS
Being assigned to build something from nothing thrusts a manager into a survival course on how to stand alone, where the premium is on initiative and courage. Such assignments can be within the home country, but they are frequently overseas and geographically isolated. They can include new plants or product lines, new markets or subsidiaries. Challenges include finding an office, building a staff out of inexperienced people, developing a market from scratch with a minimum of guidance and support from the home office, and meeting the "bottom-line" with minimal financial resources. In addition, start-up managers confront a number of adversities beyond the job itself, such as dealing with social, political, and cultural problems; creating new policies to meet the special needs of the local situation; and coping with wilderness and terrible weather. The risks can be enormous, but the rewards of creating a successful new operation from nothing can be significant for the company and a career high point for the manager. Clearly, start-up experience yields the purest opportunity for leadership development.

TURN-AROUND ASSIGNMENTS
Normally, in any business activity most operations function smoothly and, it is hoped, profitably with a certain amount of fine-tuning by management. However, things don't always work that way. A division

can get into trouble, either financially or operationally; scandal and fraud may be discovered; serious personnel problems can surface, all calling for immediate action to "fix it." The mandate may be to re-organize, clean house, cut costs, control inventories, overhaul purchas-ing, install new systems—do whatever it takes to turn the situation around. Usually, the fix-it manager has the authority to do the job; but not always, as in the case involving multiple divisions or functions.

The challenge of a turn-around assignment is not only to tear some-thing down but to effectively rebuild an organization, and to get people to work well together toward a changed and improved operation. Put-ting something back together is quite different from setting something up which is new and right in the first place.

LARGER SCOPE ASSIGNMENTS
Taking on more responsibility is less traumatic than start-up or fix-it positions, but nonetheless such assignments are also highly develop-mental. Larger scope jobs usually fall into three categories: promotions in the same function or area; promotions in a different function or area; and rotational or lateral moves. The greater the leap in scope, the greater the learning opportunities. For these assignments, the need to develop subordinates and teamwork is crucial, because newness and unfamiliarity will not permit a manager to do it alone. Self-reliance must give way to dependence on others to achieve objectives. Executive and leadership skills override functional ability, which is a key lesson underlying larger scope assignments.

Developmental line assignments involve change and discomfort; the greater the change, the greater the opportunity for learning. Managers with these assignments have to learn to survive, and as result acquire the new competencies so necessary to future career success.

Role of Bosses and Others

Successful assignment performance demands the skills of motivation, leadership, persuasion, and negotiation to bring about the desired re-sults. Learning from significant experiential assignments can be best characterized as getting the job done—pursuing and reaching a specific

goal that frequently involves new business and technical content—
through effective interaction with a number of people. On the other
hand, another kind of learning comes from interaction with a specific
person, an organizational superior or boss. An immediate superior can
have an enormous influence on the attitudes, habits, and expectations of
the subordinate. Their relationship can be a learning event in itself.

Bosses come in a variety of sizes, shapes, and managerial styles.
Some are good, some are poor, and some are a combination of qualities,
but all will have an emotional impact that, in some cases, can lead to
lasting change in the development of the subordinate manager. Seen in
another way, bosses can serve as a special substitute for direct experience, as there may not be enough assignments to go around or there may
be learning gaps in such assignments. Bosses are role models. The
subordinate manager observes how they treat subordinates, how they
make decisions, and how they espouse the values of the organization.
These observations can provide a highly developmental experience. An
exemplary boss can demonstrate how to do things correctly, effectively,
ethically, and supportively. On the other hand, an intolerable boss provides negative learning. The lessons here may be what not to do, how
not to behave, how not to cope with stress, and so on.

The relationship between subordinate and superior has been a subject
of fascinating interest throughout the centuries. For instance, Machiavelli in *The Prince* (1513) wrote as follows:

> When ever the minister thinks more of himself than of the prince, and that in
> all his doings he seeks his own advantage more than that of the state, then the
> prince may be sure that the man will never be a good minister, and is not to
> be trusted. (Lerner, 1960)

Machiavelli's concern about the behavior of ministers toward their
prince points out another set of principles underlying the relationship of
subordinates and their bosses applicable in the modern world. Dr. Eugene Jennings (1980), of Michigan State University, proposes that effective functioning in today's managerial world demands a deep understanding of the informal organization and its power structure. The
concept of sensitivity to the needs of one's boss is central to Jenning's
research. His priorities must take precedence over yours, and they must
be understood even though they might not be stated explicitly. Cues
such as "I wonder what the current morale situation is at plant 'X' "

should suggest the need for investigation of recent opinion survey data and selected contacts with key executives at that plant, followed by a brief report and recommendations.

To make the boss look good, never act smarter than he is in an open meeting with his peers, for example; to do so has led to the end of otherwise brilliant managerial careers. Never be too busy with your own projects to drop everything and respond to an immediate request for help from your boss. A crucial subordinate can flesh out and complement the boss's areas of weakness in such a way as to enhance the entire operation. Sensitivity to his needs and support of his objectives can be achieved without the loss of identity and personal competence if done properly, and can lead to an important component of the development process.

You can also learn from subordinates and peers in somewhat different ways, from subordinate roles in large projects, and peer working relationships in task force assignments. But what you can learn from bosses is somewhat unique: management values; human values; executive styles; and politics.

4

Succession and Career Planning

Succession Planning

To guarantee survival and success in dealing with our rapidly changing environment, corporate strategic planning must include all of the resources required to reach corporate objectives. The quantity and quality of human resources, particularly executive resources, are critical and must be planned for. As put so well by Peter Drucker:

> The question of tomorrow's management, is above all, a concern of our society. Let me put it bluntly—we have reached a point where we simply will not be able to tolerate as a country, as a society, as a government, the danger that any one of our major companies will decline or collapse because it has not made adequate provisions for management succession.

Executive succession cannot be left to chance on the assumption that "the cream will rise to the top." According to a number of recent studies, an assured supply of managerial talent has been one of the main concerns of chief executive officers. Executive continuity systems are in place in a significant number of large, well-known, successful companies, and are being installed in an increasing number of other firms, large and small. In such systems, the entire organization is committed to the program. The corporate office regularly reviews business-unit executive succession plans. Small but effective staffs keep the system going, including a dedication to and use of formal executive education programs, both internal and external.

Succession planning systems are known by a variety of titles: Executive Continuity, Executive Resources, Executive Manpower Systems, Executive Development, Continuity Programs, to name a few. All of these systems deal with the identification, development, and upgrading of managerial personnel.

The term *executive* in these systems refers to individuals occupying critical positions, including top offices in the organization and those near the top. They take into account current and projected executive positions in that particular organization. Categories for succession planning tables typically include

- Incumbent executive (key, top positions)
- Replacement executives for these positions
- High-management potential (HMPs), younger managers with potential to reach top positions

Openings in executive or critical positions periodically occur, sometimes unexpectedly, and succession planning assures that a qualified individual is ready to step in and take over. Thus, there is a well-orchestrated plan to assure continuity of qualified people in the executive ranks, and with the solid support of the CEO and corporate office, uncertainties about the long-term quantity and quality of corporate leadership should be minimal.

Successful succession planning systems are built on the premise that recruiting and selecting the right kind of people is crucial. Promotion-from-within policies provide an internal talent pool, and ensure that young managers can see growth and opportunity within the system. It is hard to imagine that a successful system could exist under "revolving-door" conditions where many executive positions are filled from outside with the help of executive search firms, indicating that an internal talent pool doesn't exist or is of poor quality.

While succession planning systems can provide opportunities for career advancement, it should be pointed out that succession planning is different from career planning, in that succession planning is a corporate responsibility. The company's job is to provide opportunity and performance feedback, not plan careers, which is the responsibility of the individual. In brief, companies plan succession—people plan careers!

Career Planning

What is a career? According to *Webster's Encyclopedic Dictionary,* a *career* is a specific course of action or occupation forming the object of your life. It is what you want to be or do over a forty-year span, usually in the framework of an occupation. Few children decide what they want to be and end up that way. Serious career concerns begin to surface during high school and start crystallizing in college when the choice of a major course of study is made. Careers in professions such as law, medicine, education, and science require graduate work in the chosen field prior to full-time occupation. In the managerial area, which is currently thought of as a profession, an advanced degree (MBA) can be valuable in preparing an individual for an executive career in business. Such credentialling helps achieve entry into the corporate environment but does not assure success, which is dependent on performance, not degrees.

As has been previously stated, career planning is the responsibility of the individual, not of the corporation. The company does succession planning by providing opportunity, feedback, and support in order to make the best possible use of its executive resources. The individual, on the other hand, must operate from the perspective of enlightened self-interest. According to Peter Drucker, the way to look at careers is to ask, "What can it do for me in what I want to accomplish?" To answer this question you must know what your career objectives are and pursue a carefully thought out planning process to integrate goals and opportunities.

Career options include single-company careers, multiple-company careers, single-functional or multifunctional careers, and private and public sector careers among others. Single-company affiliations have been characteristic of a number of large, successful corporations in the past, including AT&T, IBM, GE, and many others, where promotion-from-within policies have been consistent with managerial strength, effective succession planning systems, and corporate success. However, trends indicate a move to multiple-company careers and the diminution of company loyalty as a basis for growth and development. Another Drucker observation states: "You owe no loyalty to your employer other than not to betray secrets. Be ruthless about finding where you belong." Trends previously mentioned show increased intercompany mobility of executive talent, particularly in the international business environment.

Making career decisions involves a number of trade-offs among personal needs, family needs, and organizational needs. The changing characteristics of the family in the United States can have a significant impact on executive career choices, mobility priorities, and corporate policy.

- One-half of marriages end in divorce.
- 40 percent of children live in a one-parent home.
- One-third of women work outside the home.
- Dual career families are on the increase.

Today an individual can experience a number of career changes in a lifetime, with second, third, and even fourth careers not being uncommon. Major changes in functional affiliation can offer highly developmental opportunities, such as the shifting from engineering to marketing in some companies brought about by downsizing and the requisite manpower redeployment. At the more senior management levels, it is not unusual to see executives take an early retirement and accept positions as dean of a major business school. Others will leave business and take on public sector posts in local, state, or federal government.

What we are seeing is increased mobility on an inter-, as well as intra-, institutional level, with executives broadening horizons and seeking fulfillment in a wide variety of ways. Change and challenge seems to be the name of the game in a world of growing complexity and interdependence, thus calling for new levels of executive flexibility and competence.

Mobility and the Development Process

The relationship between mobility and competence has been the subject of a great deal of research by Eugene Jennings and many others in the field of executive development. As Jennings put it:

> It has been said that competence leads to mobility . . . where in actual fact, it is quite the reverse, that mobility leads to competence.

The findings may seem paradoxical in that exemplary performance in a given job may not necessarily prepare a manager for greater responsibilities. Even if the current job grows in complexity and the boundaries widen, the repetitive nature of dealing effectively with a known set of requirements may be counterproductive in terms of growth and development.

Growth and development require change and new challenges, which are characteristic of highly mobile career paths. Taking on new assignments at regular intervals and performing successfully leads to a growth in self-confidence so necessary to the achievement of long-term career goals. And, having mastered the first new assignment, the next one will be that much easier to deal with. Thus, a new kind of competence comes in to play, that is, the ability to confront the unknown, to effectively motivate strangers, to understand a new set of priorities, and to obtain desired results.

The traditional routes to top management positions are through what is known as functional channel mobility. That is, the senior corporate officers make it to the top by way of one particular function, such as marketing, finance, manufacturing, or engineering. For example, in DuPont, the engineering route prevails; in IBM, marketing dominates; at GM, it's usually finance or manufacturing. Depending on the corporation, all other functions terminate at a somewhat lower level. Historically, if an executive is moved out of the function in which he had early success, especially into staff work, it is a sign of failure.

Today, and into the future, we see a changed organizational environment characterized by rapid growth; smaller, leaner structures; and downsizing and flexibility. These characteristics require crossover mobility whereby a manager may have a number of different functional experiences, moving from engineering into marketing or finance, with such moves placing a premium on managerial rather than on technical competence. As stated in Chapter 3, a combination of line, staff, task, assistant-to, and international assignments will more likely be the norm in the successful executive career. Assignments might then also involve functional shifts as well as cross-divisional tracks. Hence, mobility patterns in the future will have both horizontal and vertical dimensions, demanding a higher order of competence than that of the traditional organizations of the past.

Mentors

It is an accepted historical fact that a young person learns a trade best when studying with a master. Hence apprenticeship has been a requirement to career success in the arts and trades. The applicability of this master-apprentice concept to business has not been recognized until recently. A number of studies of the mentoring process in business and industry have been conducted over the past several years, including one conducted by Gerard R. Roche (1979), former president and chief executive officer of Heidrick and Struggles, a leading management consulting firm. The results of this and other research show a high correlation between successful executives and mentors, whereby those who had mentors received a higher compensation at a younger age and were happier in their careers than executives who did not.

There also seems to be a correlation between career planning and mentoring. Over the years covered by such studies, it was shown that more executives who had mentors followed a career plan than those who did not, and that the combination of career planning and mentoring probably accounts for the higher compensation for those executives who had a mentor. This concept has its greatest applicability to younger managers on their way up, because the first fifteen years of their career is the time for learning and growing when mentors can have the greatest influence. Mentors are particularly important to minority-group and women managers who are serious about doing so achieving higher management levels.

In business, a mentor is usually someone higher up in the corporation who can have a favorable influence on the career of a younger manager. The mentor need not be an immediate superior although it is so in a number of cases. The mentor can provide support and assistance in a number of ways, particularly through the sharing of knowledge and experience of the organization and its power structure. He or she can provide perspective and counsel on career paths: on who the important people are and how to get exposure to them: on strategies for ongoing growth and development; and on an understanding of the informal organization, which is so necessary to effective career management.

Executives who have had mentors are most likely to become mentors themselves, apparently having discovered the importance of this special kind of support for younger managers coming up through the organization. The resurgence of mentoring by the executive group aged 55 and older who are approaching retirement reflects an apparent desire to pass along what they have learned. At this stage, career goals have been largely met, and the contribution of these senior people can take the form of teacher, counselor, and advisor not only to those reporting directly to them, but to any younger protege who hopes to reach the higher levels of the organization and make a lasting contribution to the continuing success of the company.

The most telling difference between those managers who have had mentors and those who have not is in terms of job satisfaction. Mentored executives report greater pleasure from their work. Since "work" plays a central role in the lives of executives, it should be enjoyed!

Criteria for Success

What advice should be given to talented young managers planning their careers? Here are some thoughts we gleaned from successful executives on this subject:

Be prepared for change. It is the rule, not the exception, so take advantage of it. Seek out expanding, broadening opportunities. Don't look for comfortable jobs that rest on technical skills. You've got to put yourself under pressure.

Watch out for complacency.

Be flexible in taking jobs and geographic moves. Accept challenges as they come.

Be a Quick Study

We usually think about the characteristics of high-management potential employees (HMPs) on succession planning tables when considering successful career strategies. Depending on business conditions, corporate cultures, and other variables affecting opportunity, the fast-track HMP is worthy of study.

HMPs can be defined as those making twice their age in salary, the so-called early arrivals. If they are competent and quick to learn their way around, they will get the new job under control quickly, make a significant contribution, and move on. According to Eugene Jennings (1980), you can learn all that is of importance and get on top of a new assignment within two years, and then move on. Here the issue is to figure out which 20 percent of activity counts 80 percent in performance evaluation, an exercise in intelligent prioritizing. As Jennings puts it, you have to obtain the loyalty and support of the "natives who can win over the immigrants" and obtain from them the territory's roadmap, that is, the who, what, values, and so on in the new environment. Only in this way, can a young manager on a fast track move his or her career forward.

HMPs are careful to avoid dead-end jobs, particularly if they are minorities or women. Potentially stagnant positions are frequently staff, with titles such as Urban Affairs Officer, Affirmative Action Coordinator, Special Projects, and the like. If such a staff assignment comes up, it can be used for a learning experience, but specific plans for exit must be agreed upon.

Learn from Failure

Although we deal in some depth with the issue of derailing and failure in chapters 5 and 6, a few comments here about business mistakes are in order. Most successful executives will admit to having made at least one (sometimes several) serious business mistakes in their careers, including costly ideas that didn't work out, failure to take advantage of opportunities, and conflicts that got out of hand. Significantly, experience has shown that the leading and underlying cause of business mistakes is some kind of failure to deal properly with other people, such as underestimating the importance of other people to the success of a project or an idea. Mishandling relationships with subordinates, peers, and superiors must be recognized and corrected to assure recovery. To do so requires a critical evaluation of self and of one's sensitivity and management style.

Failures and mistakes can be viewed as positive or negative experiences. For example, a young HMP who "blows it" could have the event evaluated as "developmental," a contribution to learning, because he or she is well thought of by management and seen as having great potential. On the other hand, another manager who is not thought well of and who "goofs" may be evaluated as having made a "mistake," and thus being incompetent.

Regardless of their causes or evaluations as positive or negative events, the successful manager views mistakes and temporary failures as learning experiences and opportunities for personal growth.

Be Ready to Relocate

Successful careers frequently involve physical moves. If we track high-potential managers in leading firms, it becomes clear that career success is closely related to taking any opportunity that is offered, and many of these opportunities involve relocation to another part of the country. You may pass up such an offer once, but the chances of another developmental opportunity being offered may be remote. Of course, relocation can be a burden for the family; houses must be sold, affairs must be put in order, and a number of other sacrifices must be made. But such is the price to be paid to move a career forward. Some companies recognize the costs of such moves, to the firm in terms of high moving and living expense, as well as to the individual and his or her family. Policies have been formulated whereby promotional opportunities should be found within a given geographical region wherever possible to minimize physical moves. In any case, considering relocation frequently comes down to an individual decision about the trade-offs for career success and family requirements.

While relocation often provides an opportunity for advancement, you should also consider the frequency of job changes. We have stated as a general rule of thumb that the time spent in one position should be sufficient to get on top of what is required, and that period of time should be adequate to exhaust most of the learning opportunities therein. In one major corporation, a study was conducted on mobility patterns of the top twenty executives. The results showed that the median stay in any position was 2.1 years, with the maximum tenure being 2.9 years, and the minimum occupancy being 1.2 years. The issue may be one of too fast or too slow. If too fast, burnout could occur or significant failure could jeopardize future career success. If too slow, there might be a loss of potential contribution and reduced effectiveness in use of important executive resources. There is, of course, a balance to optimize growth and challenge, which, naturally, depends on business conditions, growth patterns of the industry, the company culture, and other variables.

Prepare for Uncertainty

Mergers, cutbacks, reductions at the middle-management level, and layoffs are realities that point to the need for a strategy of dealing with such uncertainties to ensure a successful career. The president of Retail Recruiters International conducted a survey of client companies showing that the average employee who was fired is a 47-year-old middle manager who had been with the company for more than ten years, and hadn't been promoted in five. It is time to move on if one of the following situations applies:

- You have been working at the same job for more than five years and haven't been promoted.
- You are not contributing anything new.
- You are not learning anything new.
- The company is planning to be merged or sold.

Making a move under these conditions is not job-hopping, rather it is realistically dealing with the current and future business environment. Employees who are comers are also movers, and if any of the preceding conditions apply, then you should be prepared to "move upward and onward" to where the opportunities exist.

Manage Disappointment

While no one is immune to disappointment, people who want power and responsibility are particularly vulnerable to situations where reality does not conform to their wishes or intentions. But far from disappointment being a prelude to continued failure in their careers, these episodes may really be occasions for accelerated growth and even the beginnings of truly outstanding performance.

According to Abraham Zaleznik, noted behavioral scientist, leaders from both public and private sectors have had difficulties in coping with demands and stresses of responsibility. The president of the United States has little choice in the problems that arise or even how to deal with them. Key business leaders deal the same dilemma: facing the gap between what one wants to do and the practical possibilities for action. Personal goals often take a back seat to the realities of any given situation. If those goals are unrealistic, then major disappointment is certain.

No one is immune to encounters with disappointment. As with all matters of personal development, the outcome depends on the quality of the individual, the courage which can be mobilized, and the ability for constructive introspection. With self-examination, you might recognize the impossibility of certain goals and desires, but at the same time discover new and uncharted possibilities for productive work and pleasure. When planning for a successful career, expect some disappointments, but learn to make the best of them.

5

Career Problems: Plateauing, Derailing, and Obsolescence

> If we are to support adult development on a wider scale, we will have to modify the social institutions that shape our lives. Industry and other work organizations, government, higher education, religion and family . . . all of these must take account of the changing needs of adults in different areas and development periods. What is helpful in one era may not be in another.
>
> HARRY LEVINSON

Plateauing

In every executive career, there comes a point where upward mobility ceases and further progress to higher levels of management cannot be achieved. In some cases, this plateauing occurs at the CEO level (no problem), but in most situations, we see peaking out at middle or upper-middle management positions. Three facts about organizational plateauing are clear. First, it is unavoidable; second, it is usually painful; and, third, the individual changes as a result of experiencing it.

This phenomena is usually associated with the middle-aged manager and the so-called mid-life crisis, according to Levinson and other social scientists. Most men will come nowhere near doing what they want to do with their lives, resulting in some degree of difficulty and frustration as a result. After age 45, they turn gradually to matters outside of themselves, may be increasingly concerned with ideals and causes, and may also become more concerned with finding a purpose in life. These shifts in attitudes also relate to the inevitability of the aging process, which affects managers and executives as well as everyone else. Within

39

the organizational context, the increasingly important issue of the plateaued executive needs to be examined and evaluated for ways of dealing with it from both an individual perspective as well as from an organizational viewpoint.

Plateauing Due to Inability or Incompetence

Sometimes managers find themselves in positions that demand more than they can deliver, resulting in marginal performance. Such a situation causes enormous difficulties for both the individual and the organization. Loss of confidence, stress, and a depressed self-image can adversely affect an already low level of performance. What can be done in such cases? Demotion may make most sense whereby the individual is placed in a less demanding position in line or staff, where his or her abilities can be effectively utilized. Reassignment together with appropriate counseling can help assure a continuing contribution to the success of the business.

Depending on the particular situation, firing may be the only appropriate action to be taken. If carefully handled with generous financial arrangements and assistance in finding another position, release from the business and, perhaps, an intolerable situation may prove to be blessing in disguise for both parties. There have been many cases where a fired executive thanked the company for taking such an action, which resulted in a new and productive career in another organization.

Other courses of action include tolerating poor performance by off-loading certain responsibilities to others; that is, managing around the individual by minimizing the areas of weakness and preserving the areas of strength. This approach is a form of isolation that, in certain circumstances, may make sense to the individual and to the company.

It is hoped plateauing due to inability and incompetence will be increasingly rare because enlightened firms will pay more attention to promotion policies and individual competence profiles. Regular and sensitively handled performance reviews should help assure that there will be an appropriate match between job requirements and individual abilities.

Plateauing Due to Lack of Opportunity

The traditional organizational structure in business has been hierarchical in nature, with up to ten levels of management between the line supervisor and the chief executive officer (CEO). As you move up, fewer and fewer positions to aspire to are available. However, there is a trend in organizational structure to move away from the traditional pyramidal design to flat, decentralized forms with reduced managerial levels. This trend suggests that the opportunity issue may take on new dimensions in the future.

The most frequent cases of plateauing occur within a traditional structure, whereby an otherwise competent executive seems stalled in terms of promotion and career progress, resulting in a loss of initiative and morale. The real or perceived lack of opportunity may be due to factors other than the structural issue. The company may be in a low growth situation within its industry necessitating a reduction of promotional opportunity. Or there may be threats of mergers and takeovers that can lead to a holding pattern and uncertainty. In addition, the promotion channels may be blocked by an inordinate number of senior officials who are holding on until retirement. An increasing number of companies, recognizing the latter condition, provide out-placement or "window" plans for senior managers, and thus open up promotional channels.

Other courses of action include special assignments, task force work, job enlargement, and participation in executive education programs. The challenge is to keep these good people. Their morale and motivation can be maintained through other avenues of contribution. As previously mentioned, the older or middle-aged manager may have to face a change in career focus: moving from a single-minded pursuit of top management positions to a recognition that responsibility to self and to the business may have to be redirected to the development of subordinates and younger colleagues. A new role as a mentor and teacher, sharing long experience with others, can be enormously rewarding as a new career goal. Apart from leaving the firm for greener pastures, a reexamination of abilities and life goals may lead to a new maturity and a heightened sense of satisfaction.

Derailing

What is derailing? Unlike plateauing whereby managers find themselves promoted beyond their abilities or to positions in which opportunity for further promotion is apparently nonexistent, derailing is self-inflicted. It occurs when a manager of promise is expected to rise in the organization, and who wants to advance, is fired, demoted, or stalled in mid-career for reasons of his or her own making. According to the Center for Creative Leadership (CCL; McCall and Lombardo, 1988), the principle reasons for derailing are as follows:

- Difficulty in molding a staff, cronyism; being unable to resolve subordinate conflict.
- Difficulty in making strategic transitions, unable to adapt; mired in tactical detail.
- Lack of follow-through, incomplete jobs and leaving people dangling with unmet promises.
- Poor treatment of others, overambition and isolation.
- Overdependence, staying with the same boss too long or relying too much on one skill.
- Disagreement with top management, inability to sell a position or inability to adapt to a boss with a different style.

Derailing can be associated with the career crisis syndrome. According to Jennings (1980), studies of men experiencing career crisis have shown the presence of a rather extended period of previous administrative difficulties that have not been successfully resolved and that, as a result, bring about a consciously felt career crisis. In many cases, these past difficulties have involved interpersonal relations, where little or no feedback has been given and greater weight has been given to his or her accomplishments rather than to flagging abrasiveness and lack of sensitivity. In many such cases, the manager has been recommended for promotion and the problem thus passed on to others. Put another way, few high potential managers derail because they lack technical skills or fail to get results. They fail because they focus too much on their strengths and ignore their flaws, or won't admit they have any.

For many managers facing such a career crisis, the underlying problem seems to be that their ability to learn from previous experience remains largely unfocused. Bright people do not always figure things out for themselves, and research shows that even among the best who reach the top echelons of their organizations, perhaps only one half could boast accurate self-awareness.

The costs of derailing to both the organization and the individual can be significant: exit costs, wasted training and education, hiring and restart costs, the cascade effect of one empty slot at the top leading to multiple position shuffles, and the damaging effect on the manager in terms of morale and loss of self-confidence. According to Jennings (1980), the "unconfidence effect" leads to a heightened tendency to filter, that is, seeing things as you want to, not as they really are. This further reduction of effectiveness can be cumulative, resulting in fear, anxiety, guilt, shame, and frustration. Confidence must be restored and self-awareness must be introduced for an accurate assessment of the real causes for the derailment. Once understood, rescue operations can commence with the hope that the individual can be saved, a career restored, and corporate loss can be minimized.

Derailing, then, is involuntary and punitive, and is not an infrequent occurrence. According to a number of studies conducted by CCL (Mc-Call and Lombardo, 1988) failure rates regularly occur in corporations, in some cases as high as 33 percent for senior executives, with "hardship experiences" resulting from missed promotions, firings, or business failures having been suffered by almost all executives in a number of leading Fortune 500 companies. Armed with this information, the wise executive can formulate plans to avoid derailing or, barring its prevention, initiate rescue operations.

Prevention of Derailing

Self-perception and self-confidence seem to be integral parts of both successful and unsuccessful careers. According to Deal and Kennedy (1982) in their publication, "Corporate Cultures," how you perceive yourself vis-a-vis your peers can have a strong influence. Their "Dummy Theorem" states that in any group of people, a certain number of them will be dummies and the rest nondummies. The theory further states that with any peer group, people categorize themselves as either a dummy or nondummy based on their perception of themselves. An example of the influence of self-perception involves the philosophy of the Mitsubishi company concerning middle managers who misperform (in an environment of lifetime employment). If the reason for the lack of performance is not clear, then the manager is promoted. The chairman of Mitsubishi states that in 72 percent of the cases of such promotions, the manager's performance immediately improves because the promotion improved his perception of himself in relation to his peers. While not solely the answer to the derailing phenomon, positive self-perception can be a powerful motivating force in the complex equation of executive career success.

Probably one of the most important aspects in the prevention of derailing is the proper fit of an individual to the job. Careful selection of next assignments should be central to any succession planning system. A detailed evaluation of an executive's performance, including strengths and weaknesses, should be thoroughly done, so that when matched with the profile of requirements of the next position, a reasonable fit is achieved and the chances for success are maximized. If, for example, an individual seems to require frequent contact with and support of superiors, he or she should not be given a regional marketing manager's job requiring travel, independence, self-reliance, and little regular interface with superiors. Such a match could be a recipe for failure. Thus, a properly managed selection process can help assure that executive derailing will be minimized.

Perhaps the most elaborate technique for assessing the potential of lower and middle managers and thus reducing the chances for derailing involves the use of assessment centers. Utilizing standardized exercises, common criteria, and multiple raters who are often high-level executives, these centers have a fairly consistent track record in predicting success. The concept is that by participating in the activities at such a center, a manager is exposed to a simulated managerial environment, demanding the competence and performance required on the job, but in a low-risk situation. While the initial purpose of assessment centers was to aid in the management selection process, the most productive aspect has turned out to be the enhancement of individual development through feedback from observers and peers. Such a microcosm of reality emphasizes managerial rather than technical skills, and so can point out areas of strength and weakness in interpersonal behavior, communications, priority setting, handling pressure, and dealing with ambiguity. Despite the fact that assessment centers are labor intensive, costly, and are not "real life," their use can provide yet another avenue of feedback for the individual manager—feedback that, if taken seriously, can be instrumental in the prevention of derailing (as well as being helpful in succession planning).

While the formal performance appraisal systems used in most organizations are subject to controversy, it should be noted here that regular, open communication from superiors, peers, and subordinates on your performance provides a critical element of perspective on how you are doing. Strengths are recognized, successes identified; but, most important, areas of needed improvement can be brought out into the open for examination and rectification. Perhaps we can learn more from our mistakes than from our successes. Perhaps the question we should seek to answer is not "how good am I?" but "what can I do to improve?"

Rescue of Derailed Executives

There are two players in the rescue process: the derailed manager and the organization. The derailed manager has to undergo significant changes, learning to calm down, to admit errors, and to work at new and more effective management styles. Most derailed executives focus so narrowly on performance and "hard" management skills that they do not reflect on the broader lessons of experience. Hence, organizational intervention is necessary, which usually involves removal from the current assignment, feedback, course work, counseling, coaching, and a series of new assignments designed to generate a broader set of leadership skills. Above all, the individual has to take ownership of the problems that got him or her in trouble.

One of the most effective approaches of rescue is through targeted multiple assignments; that is, a blending of jobs and course work designed to reinforce new and more productive behaviors. This approach could include a combination of workshops in listening skills and group process interwoven with specially selected assignments under role-model bosses who provide regular feedback on performance. If and when significant change has taken place, management should take an active role in placing the executive back on the track. In many cases, this move would involve a lateral transfer to a new division that offers new and different challenges and opportunities, and where new skills can be put to best advantage.

The cost to the company in a particular rescue case might be $10,000 and having an executive two months away from the job in various courses and workshops. However, if only even half of the derailed executives can be saved, the investment is well worthwhile when compared to the cost of replacing a key member of the management team.

Obsolescence and Age

To understand the connection between obsolescence and age in the executive group, it might be helpful to take a look at this phenomenon in the engineering fraternity. We have previously stated that the half life of an engineer is four years, that is, one-half of what was learned in college is obsolete in four years. According to one recent study, the performance of engineers peaks in the middle to late thirties, and there is a definite trend toward earlier obsolescence; the years of high performance are starting and ending sooner. This obsolescence is the result of a growing technological sophistication in business and industry, as well as rigid performance appraisal systems, inequitable job assignments, and insensitivity to the needs of older professionals.

Does this mean that older engineers must necessarily be less competent? Less responsible? Less creative? Not necessarily. As one example out of many, Shakespeare wrote comedies and sonnets until he was forty; all of his major tragedies were written after this age. A person's creativity changes as he or she grows older and so does his or her needs. It is important to work with some of these changes instead of against them.

How does this phenomenon of obsolescence and age apply to the managerial and executive group? It seems clear that there is a trend toward older managers at the middle and upper-middle management levels. An example from IBM indicates that the average age of high-potential middle managers attending the internal Advanced Management Program has moved from 31 years old in the late 1960s to 37 or 38 years of age in recent years. At the senior executive levels, mandatory CEO retirement at age 60 is a policy of some companies, while others do not limit CEO tenure by any age constraint. The increasing use of "window" plans has achieved some reduction of older professionals and managers in many corporations.* However, the projections of workforce aging for the next twenty years strongly suggest that there will be a significant impact on the source of and profile of managerial personnel in the future, as well as a decrease in middle management levels through restructuring and downsizing. Hence, middle managers will probably be fewer, older, and necessarily more competent than their predecessors.

What can be done, then, in view of the inevitability of an aging workforce and increasing demands for state-of-the-art technological sophistication? Obviously, continuing education is essential to prevent

*"Window" plans offer an attractive financial incentive for early retirement and the decision to participate must be made within a designated time period, or "window."

obsolescence. Such education is a responsibility of both the individual and the organization. This approach also suggests the need for sabbatical leaves for three to six months rather than attendance in part-time courses to get on top of new technologies. Better management practices include assignments to new-technology projects that stretch and challenge rather than to existing old-technology projects. Organizations must face the fact of retaining and effectively utilizing a smaller, older, and more competent workforce to stay competitive.

As defined in *Webster's Encyclopedic Dictionary,* obsolescence is the state or process of becoming obsolete, that is, disused, neglected, out of fashion. If we assume that there is no necessary correlation between age and obsolescence, it may be well to explore the managerial obsolescence issue in today's and tomorrow's organizations. An executive's effectiveness can be roughly evaluated in terms of knowledge, skill, and attitude requirements of the job, the combination of these qualities which we have defined as executive competencies.

In Chapter 2, a projection of executive competencies for the twenty-first century was formulated into five categories: environmental, leadership, managerial, interpersonal, and business. To prevent obsolescence, these abilities must be enhanced if they are present, and must be acquired if they are lacking. As we have seen, meaningful on-the-job experience can play a vital role in the development of these required competencies. Leadership skills can be learned or strengthened through experience by conscientious feedback from bosses, and even by temporary derailment. Negotiation competence can be strengthened through task force assignments and management simulations. And, above all, continuous, lifelong education through regularly scheduled executive development programs, both internal and external, will provide an excellent hedge against obsolescence. Personal and professional growth must be a primary objective of both the individual and the organization. Such growth requires an investment of time, energy, and resources to help avoid obsolescence in the executive group in an increasingly complex and demanding world business environment.

Part II

Lifelong Learning

6

Success Versus Failure

> The secret of success in life is for a man to be ready for his
> opportunity when it comes.
>
> BENJAMIN DISRAELI

Personal and Professional Philosophy

Man is a political and social animal. His success or failure in his
organizations is largely determined not by his level of professional
knowledge or intellectual and analytical prowess, but by how he in-
teracts with other people. He must understand the interests, strengths,
and weaknesses of others. He must know how to work with those he
disagrees with, to speak his mind and state his position without offend-
ing others unintentionally. He must know who his friends and enemies
are. In short, he must be aware of his animate environment, of its
personal dimensions; for any environment is a complex of interacting
human forces that need to be led or governed.

Whether in a business or political organization, to accomplish one's
managerial mission, effective means that are morally acceptable are
prerequisite to any hope of success, just as being ignorant of such
effective means is sure to lead to disaster.

Managerial and executive positions are authoritative in nature, but,
more important, are powerful, a fact that tends to make people uncomfortable or even fearful. Power is morally neutral, and can be used for
good or evil. The key to success is to understand it and use it wisely. By
handling power realistically, a manager or executive does not necessarily discard virtue.

Machiavelli (Lerner, 1960) offers some further thoughts on the
achievement of success. On gaining a reputation, he says "do glorious

51

things, have significant friends and enemies, love, merit and honor the able, reward the productive, and mingle with the people occasionally." On the perceived gap between "seeming" and "being," the following questions can be raised: Is it immoral not to reveal all oneself? Is it wrong not to speak the truth in all circumstances, especially when others will surely be hurt? Is it necessarily "game playing" to adapt behavior to circumstances? On the contrary, it is probably a measure of effective leadership to be flexible.

Max Lerner (1960) in his introduction to *Machiavelli the Prince* put it this way:

> We may yet find that an effective pursuit of democratic values is possible within the scope of a strong social welfare state and an unsentimental realism about human motives.
>
> Hence, we can learn much from experience, provided we are open to its lessons. Success and failure are relative terms, not to be measured by rank or organizational level, but by being all we can be, by realistically identifying life and career goals and pursuing them energetically and with good grace.

Executive Development Strategy

In the previous chapters, we present a series of alternative futures: scenarios of the global business environment and what challenges to success corporate and other institutional leadership will probably be facing in the twenty-first century. These projections provide the basis for a related set of executive competencies required for success; that is, for effectively dealing with this global business environment. These competencies should be the focal point of executive development activity, which has two essential dimensions: on-the-job experience and formal executive education programs. In order to ensure that both of these elements are addressed, a planned effort must be made within an organization to provide its executives with the vital linkage between experience and education.

A corporate executive development strategy can provide such a linkage based on succession planning that has, as a key feature, an individual development plan (IDP) for every executive in the system. At each career step, a formal development activity that is appropriate, relevant, and timely should be in place to assure that required development actions happen when they are supposed to happen and not left to chance. The IDP should contain data on strengths, weaknesses, recent development actions, development needs, possible future assignment, and employee preferences (where the career plans of the individual are introduced).

Future assignment planning should indicate next positions for that person—line, staff, task force, assistant-to, overseas—together with time frames and readiness status. By means of this process, development needs are flagged and then can be addressed both through the careful selection of that next assignment and through a series of formal executive education programs linked to the requirements of each career point.

Formal off-the-job executive programs fall into two major classes: internal and external (see Chapter 14). A growing number of organizations are setting up their own internal executive programs to meet specific corporate development needs. Furthermore, these firms are making increasing use of external executive programs offered by major universities and other institutions to meet the broader development requirements of their executives.

An executive development strategy should call for an alternating attendance sequence starting with an appropriate internal school, followed by a selected external program. Later, that particular manager would return to a higher level internal program, followed by yet another external program experience. Ideally, this planned sequence of programs would call for regular education throughout an entire career, with such programs being carefully selected to match offerings to individual needs and job requirements at any career point. Such a strategy forces discipline into a process that has been largely unstructured in most companies. Executive program attendance can now be approached with seriousness and purpose, and can be as well-managed as any other process. With the clout and full support of the corporate office, human resources departments and executive development staffs and line management, this planned approach to executive development can help assure the investment in executive resources will pay off in long-term success of the firm.

Going to school is not merely something nice to do if you have time. It is a developmental assignment in its own right, to be well managed as a key learning experience, and should be repeated at regular intervals throughout a career. The following chapters deal with the field of executive education in depth—how it started, how it works, and what it can contribute to growth and development.

7

A Rationale for Continuing Executive Education

And one man in his time plays many parts,
His acts being seven ages

SHAKESPEARE, *As You Like It*

The Executive Education Movement

Evolution

In the late eighteenth century the Industrial Revolution transformed the face of England and was well on its way toward doing the same in Europe and the United States. Yet a person living at that time would have been astonished at what has become a hallmark of industrialized societies: the idea of the manager and of "management." During most of the nineteenth century the owner ran the business, hired clerks to keep the accounts, trained and supervised the workers, and dealt with money lenders when capital was needed. There existed only owners and workers, and the disputes between them were direct and often brutal. The concept of a hired management with its managers, executives, and staff people scarcely existed.

Thus we remind ourselves that those concepts we now take for granted have evolved only in recent times. They were a necessary response to the massive economic, technological, and social upheavals that even now continue to transform our world at a pace that we can scarcely cope with. And so it was essential that the function and means of management be invented to handle the overwhelming complexities facing enterprises in the modern world.

With the evolution of management as a new element in the structure of organizations, there came a slow but inevitable recognition that new kinds of work and competencies were required, a recognition that managers and executives constituted a new class of people with special kinds of knowledge, skills, perceptions, and aspirations. "Management" became a twentieth-century invention as significant to industrial civilization as any that we see in technology itself.

With the introduction of management and managers, university administrators began to sense that these newly conceived functions called for new combinations of knowledge and skills that had never been taught before, and about which new theories had to be developed and tested. The new competencies had the peculiar characteristic of being best developed within the empirical atmosphere of business itself, or at least in close conjunction with it. For the most part, new educational material had to be developed outside of the traditional disciplines, although it drew heavily on them. Some instructional methods were transferable, such as the case method of instruction that had originated in law school. Others had to be invented, such as computer-assisted business simulation exercises. Early in the twentieth century there was a ferment of new ideas, of experimentation, and of publication. The turn of the century saw the beginnings of what has since evolved into a widespread system of graduate schools of administration or management at universities in the United States, England, and Canada, and that has since spread throughout the developed nations of the world.

These degree-granting courses were soon accompanied by a parallel development: the advent of relatively short educational programs specifically designed for experienced executives in the mainstream of their successful careers. That idea of continuing education was not new in history; it had appeared in various forms over the centuries in the professions of law, medicine, natural sciences, and the military. But to apply it to business management and administration was historically unique. That development is a subject in itself, and the interested reader will find a brief sketch of it in the Appendix D.

Characteristics

In order to understand a complex activity such as executive education, we must consider the dominant characteristics that define it. That is what the next few chapters aim to do, by setting out a road map for the reader as follows: the stages of increasing responsibility throughout the career of an executive and what they mean for personal growth; the special conditions required for the learning adult; some of the features of subjects presented in programs; the nature of internal corporate programs; and the selection and evaluation of university-sponsored programs.

For now, you should keep several points in mind when exploring these matters. First, the activity of executive education is empirical and experimental; it is difficult to discover any unifying theory embracing management, managers, leaders, or the education and development thereof. However, there is no lack of hypotheses, and we discuss some of them later in this chapter. Perhaps that empiricism accounts for why formal executive education developed extensively first in the pragmatic climate of America. Yet we now see it spreading worldwide and gathering strength from its intercultural exchanges. It is becoming a kind of educational dimension to the binding forces of international economic and technological interdependence that seem likely to dominate the events of the next century.

Second, continuing executive education is not a highly visible activity for the average citizen, possibly for the same reasons that he or she is only dimly aware of what executives do or how business operates. Even the more traditional elements of the education establishment look a bit askance at its trade-school upstarts (the Harvard Business School, for example, is on the "wrong side" of the Charles River). The general notion is that if you are headed for a life of worldly pursuits, first you go to school, then you go to work, and that's that. The idea of a person undertaking serious studies during his or her career is a novelty in the eyes of the general public. Even managers are often ambivalent on the subject, somewhat skeptical of the whole idea.

Finally, it is worth noting that executive education programs, and the business schools that have provided most of their substance and impetus, have turned the spotlight of inquiry onto the dynamics not only of business, but onto the executive. It turns out that the executive is not a static entity in the corporate equation, but rather an ever-changing person who performs many roles during his or her lifetime. Thus, the effects of all-pervasive change seem to underscore the rationale for a lifelong learning effort by executives under the encouragement of their organizations.

The Consequence of Change

The driving force behind the growth of executive education in the twentieth century is change itself.

- Accelerating change in the economic, social, political, and technological character of the national and the international scene
- Increasing demands placed on executives to manage change rather than just to react to it
- The personal metamorphosis that a manager experiences as he advances throughout his career in the complex organizations of today

That first kind of change, the macrochanges in the nation and the world, has been considered in the earlier portions of this book that address scenarios and their effects on managerial competences. From such changes flow consequences of significance for executive educational programs.

First, we should note that the rate of change in every aspect of economically advanced nations far exceeds anything before experienced in history. A graduate in law, medicine, engineering, political or natural sciences, or any other profession is on the way to obsolescence within a decade unless a conscious effort is made to "keep up." Formal schooling plus necessarily limited personal experience are not enough. The processes of keeping up are more than most people can handle unless they have some outside help to enrich their own observations and perceptions.

In addition to the increasing rate of growth in new knowledge, just its volume and complexity are overwhelming. Everything affects everything else, and changes that once could be reasonably predicted now come unexpectedly and from odd directions. The world is like one vast power grid; an aberration in one part affects the whole.

Second, it appears that most individuals can accommodate those alarming characteristics of change while more or less adjusting their personal lives to its consequences. However, managers and executives are responsible for the on-going and future health of their organizations. And economic organizations are immediately and notoriously susceptible to the consequences of any changes in their environment. Therefore management is constantly on the firing line of impending change, trying to anticipate it, making judgments on the basis of incomplete knowledge of the facts, and acting because action cannot be postponed. Yet fundamental economic and social changes are nearly always well under way before they become evident.

Nations rise or fall depending more than ever before on how well or badly the organizations that comprise their institutions are managed. Thus executives must be held to new levels of accountability. That responsibility calls for increased levels of awareness and of excellence in performance from individuals who are entrusted with the burdens of leadership. How well they truly understand what they are doing becomes ever more crucial to the well-being of us all. In our fast changing world they cannot afford to remain ignorant. Continuing education is one way to keep up in a situation where the pace is fast, the competition alert, and the stakes high. An incremental gain in excellence can change the odds for the future success of managers and for the survival of their organizations.

Finally, personal change is inevitable within the individual who, throughout a career, moves into new and increasing responsibilities. An adult who faces a succession of new challenges goes through states of maturation as far-reaching as does a child or adolescent. In our youth-oriented society we assume that once a person reaches adulthood he or she has somehow become a fully educated, developed, and functional individual already capable of a lifetime of contribution. That may have been true for a person in a simpler age. However, for those in managerial careers today certain provisos should be recognized. For instance, young adults may be ready to pursue a managerial career, provided they have the potential— "the right stuff"—to begin with. And provided they subsequently confront and resolve a series of new challenges. And provided they continue to grow by learning not only from their own experiences but also from the world about them. A life of work and accomplishment calls for successive levels of knowledge, skills, judgement, perception, vision, and aspirations, and these are attributes that change with the years not only in degree but in kind. How they change over time establishes the nature of the person's growth, or retrogression, or transformation into a different career or a different person. The outcome must be one of those three, for the individual cannot escape some form of change. That fact is so fundamental to executive education that we must now examine it in more detail.

Stages of a Managerial Career

If you were to take a long look across the many years of a person's career in management within a highly structured institution such as business, industry, finance, government, the military, or the church, you would likely see a kind of pattern similar to what we now describe. For purposes of illustration we have to generalize it. For any one individual the shape of his or career pattern would be unique. However, for a great many individuals the relationships within their patterns would be similar.

Transitions in Perspective

Underlying all of the obvious career stages a manager goes through, we can note that he is fundamentally changing in the nature of his work and interests. His mental orientation and perspective shift from technical activity, to the tactical issues of operations, and eventually to the strategic concerns of leadership. These transitions in perspective can be described as follows:

1. During the early years on the job, a person's work is limited by whatever basic or *technical* skills he is equipped to offer. He concentrates on learning the rules of the workplace as well as on the technical content of his on-the-job training. As he accumulates experience and strengthens his knowledge and skills, he may become more influential among his co-workers, but his contributions are still those of a specialist. However, it happens that complicated tasks usually require group effort, and group effort generates the need for management and leaders. And so many an individualist finds himself gradually extending beyond his technical expertise and putting increasing attention on getting work done through others. That is an orientation to which he has given little thought and for which he is likely to have no formal preparation. Yet it is a major transition in his career, even though that might not occur to him at the time. Early successes at that point seem to be reliable indicators of more to come.

2. During the middle years he assumes broader managerial responsibilities. These new obligations become far more complex than they were before. They call on *tactical* skills and perceptions that are remote from the technical skills learned in his formal schooling or sharpened in his work as a specialist. Furthermore, his horizons of awareness now begin to extend beyond the boundaries of his own turf. He must now learn to negotiate his organizational position within a team that is concerned with broader purposes than his own. He must learn to identify with the objectives and goals of the hierarchy above him, understand the attitudes and drives of the people below him, and exchange support with his peers. He must reach new levels of self-awareness. And while all that is happening, he must begin to concern himself with business purposes themselves, with the creation of resources as well as their management, and with the establishment of objectives as well as their attainment. And so while he is learning the arts of leadership through experience, he is preparing for the third stage of perspective.

3. At this point in his career the manager (or executive) begins to see his work, his organization, and the world primarily in *strategic* terms. He must now determine where he wants to take the organization; decide when and how he is going to take it there; sell his superiors on it; and then do it. "The organization" may be the entire corporation, a product division within it, a functional cut through the entire structure or a large part of it, or a major and semiautonomous project. There are many kinds of positions at the top of organizations and many paths leading to them. Yet they share several common features: their jurisdiction is unequivocal, and within it their decisions irreversibly commit major corporate resources. It is now that the incumbent's vision, judgment, determination, political instincts, and firm grasp of reality become of paramount importance. He must formulate and instill throughout the organization a sense of common purpose. He must identify and interpret the critical economic, social, and political forces external to the organization as well as set the character and tone within it. His judgment about people must be dependable. All those tasks must fit naturally within the scope of his strategic vision.

The general case of the three stages in perspective as previously described can be illustrated as shown in Figure 7.1. This sketch suggests that the transition from one stage to the next is more or less gradual. For example, a professional engineer does not stop practice simply because she has acquired a few assistants and a managerial title.

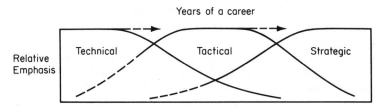

FIGURE 7.1 Stages of managerial perspective.

But as her unit grows, she devotes more attention to interactions with the functions of marketing, manufacturing, and human resources. The proportion of time spent on hands-on engineering work decreases, even though the demands on her engineering judgement may be greater than ever. As another example, an upper-middle manager may receive a special assignment that draws on his utmost capabilities and broadest vision, beyond anything he had previously experienced. An incidental purpose of the corporation in giving him that assignment could be to test him, to see how he might be expected to perform in the major leagues.

The sketch is simplified for clarity. Not shown are substages that would interlace with the broad ones. For instance, by becoming a manager of managers, the engineer faces new kinds of operating relationships which she had not experienced as a first-line supervisor. If she further advances to a position of responsibility over several key functions (such as marketing, accounting, etc.), and is subject to profit and loss measurements of performance, then she has entered into another and significant substage of management. She is no longer a junior manager but must become a generalist. She may prefer to continue thinking like an engineer, but must also start thinking like a businessperson, because doing so might be important to her continuity in gainful employment.

We all know that there is no such thing as an average person. There are only average human characteristics within given statistical sets. Likewise, there is no average career diagram that fits even one individual. Each career pattern is unique to that individual. Nevertheless, each career pattern points out the same thing: the lessons of experience gained at each stage of a career should be cumulative in their effect on an individual's future performance; an individual learns those lessons in jumps and starts; and the force of those lessons accumulate for future use at different points in a person's life. The lessons of experience, others' as well as one's own, should not be lost, but should compound like interest in the bank. Continuing executive education is designed to help make that compounding happen by stimulating the individual to think about what has been experienced or observed, and thus from that understanding gain compound interest from it.

Career Stages

We can now relate the three stages in perspective to the career stages which they underlie. The relationships are real but can be described only loosely because the terms we customarily apply (such as manager, executive, and leader, as well as organizational nomenclature) are elastic in their common usage. For instance, all executives manage but not all managers are executives. The breakthrough to executive status may be at or above the so-called middle-manager level, wherever that is. On the other hand, the term *leader* is qualitative rather than positional. A specialist may be a true leader within the sphere of his influence, whereas a key executive may not be seen as a leader because he lacks the qualities of leadership.

In spite of those semantic ambiguities, it is possible to show, as in Figure 7.2, the relationships between kinds of work (or perspectives) and kinds of jobs (or responsibilities) that a person experiences as he or she advances in an organization. Our purpose is to emphasize the fact that both managers and their work change dramatically throughout their careers. Individuals are not static; they need continual readjustment, growth, and renewal.

Figure 7.2 shows one fairly common progression of advancement in large organizations. The presentation could be sliced in other ways, and in this general concept you should allow for many exceptions. For instance, an Operations Research Specialist must often think strategically. A multibusiness general manager had better dig into the technical issues behind a plant strike. In many situations stages 4 and 5 might be combined into one organizational level. An executive in one company might be a run-of-the-mill manager in another. In Chapter 9 we consider yet another dimension of the management-thinking process, reserving it for consideration in a slightly different context.

Obviously the careers of most people do not progress through all the

| | Primary Perspective | | |
Jobs or Responsibilities	Technical	Tactical	Strategic
1. Technical specialist or professional	×		
2. First time manager or supervisor	×	×	
3. Manager of managers	×	×	
4. Manager of a key function or project	×	×	×
5. Manager with P&L accountability		×	×
6. Manager of a multibusiness division		×	×
7. Corporate level manager (group, CEO, etc.)			×

FIGURE 7.2 Seven stages of a managerial career.

various stages just described. They do not all rise through the successive layers of power, authority, or wealth as measured by worldly standards. In Figure 7.1, the dashed horizontal lines could indicate these plateaued careers. But it should not be assumed that a plateaued career is an unsuccessful one or that executive life-long education and personal growth are only for those few who rise to the top.

The idea of "success" cannot be so conveniently dismissed. Many individuals work to the end of their careers in positions that are less prestigious than those in higher echelons of their organizations or professions. They may have experienced little or no advancement during the later years of their careers, yet they feel that they have had successful and productive careers. Their aspirations and expectations have been kept in benign balance with their worldly attainments, and their associates, friends, and families share their sense of lives well lived. These fortunate and happy people! Thus we see that plateauing can mean something other than a career problem and failure as described in Chapter 5. (There is another meaning of plateau, in which the term is used to describe a level of learning, which seems to occur in a series of steps or plateaus. This phenomenon has some practical bearing in the processes of adult learning and executive education programs and is discussed in chapters 8 and 9.)

Functions of an Executive Education Program

It is easy to see, after only a moment of reflection, how quickly the stages of a career may pile upon us, like waves driving on shore from a wild sea. Many people seem to be forever caught in the troughs. But for those who ride high, the waves move fast, building up without much warning, coming one on top of the other. You don't have to be a fast-tracker to get the feeling of being rushed, of seeing the crests go by before realizing it. The old Pennsylvania Dutchman had it right when he lamented "too soon old, too late smart." Thus it is important to anticipate and prepare for the future waves of responsibility before they peak and either pass you or submerge you.

Within any segment of the managerial population there are some who early on show signs of becoming high achievers, the future leaders of their generation. These people know that their first priority must always be superior performance on the current job. But they also know that the future does not automatically take care of itself. They must keep on preparing for it, as shown in the ascending curves of Figure 7.1. These people are not so proud as to deny themselves a head start or to refuse to learn from the experiences of others, both on the job and in the classroom. But they want those lessons to be substantive and to bring new and stimulating insights to their active minds, not just confirm what they think they already know.

That attitude of demanding value out of a program experience is both the challenge and the inspiration for executive education. Executive education sessions are enhanced by the readiness of participants with enquiring minds and healthy ambition to explore matters relevant to both their present work and their future. Those are the people who are likely to benefit most from a program because they have the most to give it. However, fast movers may especially need the leavening effect of a challenging program. Their rapid success could in itself mask a flaw or deficiency in their competencies. The program might help to expose that flaw. As one perceptive student put it, "If I can get one important idea out of this course, it will be worth it." However, a bright student in a well-conducted program will get out of it more than "one good idea."

What can an executive program *not* do? It cannot substitute for experience. It can only distill the experience of the world in such a form that the student can draw from it the inferences and lessons that are important to him or her. It can impart some insight and knowledge about the purposes and techniques of management and even show something about leadership, but it cannot teach a person to be a leader. It can only stimulate the student to think hard about how leaders and management restrain or advance the fortunes of an enterprise and therefore, perhaps, to see himself or herself more clearly. Finally, it can do nothing for the unmotivated individual or the one with a closed mind.

Those limitations suggest that the process by which learning takes place is exceedingly important to the effectiveness of any executive education program. And so that is the subject of Chapter 8.

8

The Executive as a Student

> I can't no more explain
> What I don't know
> than I can return
> from where I ain't been

MOUNTAIN PHILOSOPHER

What Did You Learn?

On his way home after participating in a somewhat lengthy executive education program, John Doe was thinking about four things, in this descending order of urgency: his wife; whether his job was still there; the business crisis that occurred in his absence; and explaining to his boss what he learned.

At the office the next morning, just as his secretary was briefing him on the situation, the boss called and asked him to stop by.

"Welcome back, John. Well, how was it?"

"Oh, it was great. And thanks a lot for the opportunity. Some of the professors were just great. Some were not so hot. One or two really bombed. I guess I learned just as much from my classmates as from the classroom sessions, maybe more. A great bunch of folks!"

After a pause, John continued. "Of course I didn't buy everything. Some of it was, you know theoretical, not the way things happen in real life. And some of it didn't apply to us, because we're different." Thinking it over for a moment, John added, "But it was really worthwhile, Boss, and now I'm rarin' to get back on the job."

"Well, that's fine. I'm glad that you took the opportunity to develop yourself." Wondering what he meant, John remained silent.

"By the way," his Boss asked, "what did you learn?"

So there it was, the kind of one-two punch John had learned to expect. But he was prepared.

"Well, I'll tell you, there was so much going on, let me review my notes, and I'll come back with a report, OK?"

That kind of dialogue is common and illustrates how difficult it is for a participant to say immediately after having attended a program what was actually *learned*. His mind is a mish-mash of personal impressions and of half-recalled and disjointed flashes of insights. He can only hope that he has captured the more important ideas in his notes.

Beyond that procedural roadblock is a more fundamental one. He has tried to be a student again, after years of separation from academic-like studies. The more important things he has learned are abstract or general ideas of a kind that he has not been in the habit of formulating and expressing and exercising judgments about. And so they do not come readily to mind as he faces the boss on Monday morning.

Perceptions of Education

Three characteristics of executive education distinguish it from traditional forms of undergraduate and graduate education. First, the students in an executive program are experienced and successful practitioners in the profession they are studying about. Second, the diverse subjects in an executive program, although rooted in established academic fields of learning, nevertheless are presented in forms that stress their direct and utilitarian application to the on-going "real world" of the students. And third, the students do not come to an executive program with mental or emotional clean slates. Life has already encrusted them, for better or worse, with all kinds of things learned and unlearned, fixed perceptions, biases, and attitudes.

In all those respects the executive student differs from our perception of an inexperienced college student. Any organization that sponsors executive education should be aware of those differences because they have practical consequences for how that activity is best conducted. And they affect how the executive responds as a student, which is subject of this chapter.

The Learning Experience

Adults are conditioned to purposive recall. We are subjected to a continuous bombardment of signals, information, and ideas. In self-protection we screen out most of them and tuck others away for possible later use. That "later use" is a key step in the process of learning, for nothing has been fully learned until it is applied by the learner to some purpose of importance to him or her. Following are four instances that illustrate applications of something learned.

1. Soon after graduating from an engineering executive program one of the participants found himself traveling to Saudi Arabia as head of a team to negotiate a complex new contract. He was given a set of minimally acceptable terms and conditions to achieve or, he was told, "don't bother to return." It happened that in his executive program two days had been allocated to a study of the art of business negotiation including some lively exercises he especially enjoyed. However, he did not pay much attention at the time because he regarded contract negotiation as being far afield from his engineering work, and all a bit theoretical anyway. But after returning from his assignment a few months later, he was elated to be able to report that he had applied those ideas in negotiating better than his target terms with the Saudis, who also seemed to be pleased. Several years later it still seemed to have been a "win-win" outcome.

2. In a second example of applied theory, the participant's home base was India. The organization, of which he was the senior officer, was multiproduct and multilocation; rather loose knit; and operating under the kinds of difficult circumstances that only those in India can appreciate. There had always been a problem in getting coordination and consistent strategy and action out of its disparate marketing and manufacturing organizations. About a year after he had attended an executive program, and was ruminating over his problem, the former participant saw a way of solving it. He and his classmates had discussed things called organizational theory and organizational dynamics. As he recalled them, the subjects had been developed in the context of a hotel chain, or possibly the farm machinery industry, or perhaps a steel company that was moving into coal and shipping. He was vague about the scenarios, but the ideas he derived from them suddenly hit him clearly.

The principles and tactics applied directly to his own problem. Later he was able to report significant savings and operating effectiveness as a result of applying them. His parent company did not know or care about "organizational dynamics" as such, but it understood the improved profitability he showed.

3. Joe had always been seen as a good manager; as defined by his subordinates, he was "a good man to work for." And the productivity of his organization had usually been high. But Joe himself had often had trouble with what he called "scenes," emotional outbursts in discussions with subordinates. As he later confided, "they wouldn't listen to reason". Soon after returning from an executive program, he was confronted by one of his best people, who threatened to quit and was practically incoherent in his anger. Much to his own surprise, Joe did not try to reason with him, but merely tried to consider as best he could the feelings of his employee, while neither conceding position nor challenging that of his employee. After a while things calmed down and rational discussion became possible. At the time Joe didn't even realize he had taken a different tack with that employee; but later he did, on recalling some sessions in the program that dealt with problem-solving versus frustrated behavior, and with nondirective interviewing.

4. This final example is about an in-company four-day workshop for which the subject was "managing current assets," principally inventories and receivables. The premise was viewed from a general manager's perspective as a multifunctional issue, not just as a problem of materials-flow specialists or credit and collection people.

Each class was divided into three or four small teams from different and real businesses, each team being headed by the incumbent general manager of that business. Before coming to the class, each team had done its homework on the problems in its business and had developed ideas on how to resolve them. One of the classes was attended by the president of a major foreign subsidiary together with several of his key people. They picked up from the workshop many additional ideas about how to spot the underlying causes for inventories and receivables getting out of control and what to do about those causes.

At one point during the workshop, Dupont Charts (named for the company that developed them) were used by the instructor to analyze relationships among current and fixed assets, revenues, debt, costs, and cash flow. The people in the class were not acquainted with this particular analytical tool; they had been trained in a somewhat different approach. The program had not intended to "teach" the Dupont method; but that subsidiary team picked it up and applied it in its own business planning. It is not unusual to find in executive education that what is learned is often more a function of the student than of the curriculum or the intent of the teachers. In fact often the participants are all equally students and teachers.

Learning Reinforcement

The four examples show a progression of thought from the particular to the general, and back to the particular. The participants as a group analyze a situation or case. Then, with some help from a good instructor, they evolve conclusions or principles. Finally, they apply the principles or test them against situations in their own experience. In fact that whole progression can take place within a two-hour classroom session, since experienced adults can bring a wealth of material to enrich discussion with analogies and judgments of real-life intricacies and consequences. As the saying goes, they've "been there."

Technically that progression is called *learning reinforcement*. It is commonly practiced in all types of formal education. For executives it is especially important; to be able to act on new ideas they first have to be receptive to them. That means that they must be free of blocking tensions, most of which are part of the mental and emotional baggage they have acquired through the normal stresses of life over the years. Then they must somehow reconcile whatever new information or ideas they are learning with the information, ideas, and prejudices they have accumulated in past years, utilizing the arduous mental exercise known as reorganizational learning. Finally, they must be able to generalize—to draw out of the clutter of detail some underlying truths—and to perceive whatever conclusions or principles are relevant to the matter at hand. The capacity to generalize constructively is one of the marks of a mature mind, and is an essential attribute of any would-be leader of a complex operation.

If a potential attendee were to realize that all those processes go on in an executive program, it would be no surprise for him or her to shy away from such a den of horror. But the fact is that most of them, the more effective managers, have been engaged all along in just that kind of thinking without being aware of it. That kind of thinking has contributed to their success. So, executive students take naturally to that familiar pattern of thinking, and all their initial dread of academia, as they remember it from their sophomore days, evaporates after the first day or so of the course.

The participants also discover, like the relief of a refreshing breeze, that the executive classroom scene is a great leveler. Rank or authority of positions are acceptable only on their merits, including those of the professors. Because of their experience and maturity, the students are capable of exercising critical judgment, and they do not hesitate to do so.

Program Effectiveness

The preceding observations say that the successful executive inherently makes a "good" student. But whether being a "good" student translates into a successful learning experience in an executive program is another matter. A successful program depends on a number of factors, such as how well the program is designed to the specific needs of a particular group of attendees and their parent organizations, and then how well it is carried out.

Programs that may appear alike on paper—they all appear superb in the publicity brochures—can differ radically in the effectiveness of their educational results for different individuals or groups at different points in time. Even for any single one- or two-hour session, there will be a bell-shaped distribution curve of student reaction to it. A few will feel that it was outstanding; a few others will consider it a total loss; and the rest will distribute their opinions along some kind of curve in between those extremes. If that same session were to be given to another group there is a good chance that the distribution curve would be quite different.

Trying to forecast the success of an educational offering, like anticipating the public's response to a stage production, is a bit tricky. We can only discover through much experimenting what works or does not work; what combinations of sessions under which circumstances seem to reinforce one another; or which instructors seem to maintain the better batting averages.

Just as the "good" student is not readily categorized, the effective program cannot be mechanically diagrammed on a chart. The real program is what takes place in the mind of the student. That program is a result of a complex process of communications and energy conversion taking place throughout the whole group, but affecting each individual in his or her own way. In trying to evaluate a program or learning experience, we need to realize that we are really dealing with the mystery of human interaction, generation of thought, and intellectual invention.

The fact remains that although executives usually make good students, as we ordinarily think of that term, in practice the educational or "learning value" can vary over a wide range. We cannot confidently predict the outcome for any one student in a given instance. But we can speculate about averages based on considerable observations of what goes on in the education process. Some of the dynamics that influence the quality of the learning experience have to do with the program; others with its settings; others with the faculty; and still others with the students themselves.

Student Attitude

"Well, here I am. Now learn me!" Program directors are not outrightly challenged as crudely as that. Yet every director has noted on occasion an undercurrent of attitude among a few attendees that, in going beyond mere indifference or skepticism, borders on open resistance or anger that can be disruptive and may be infectious. At the least it encourages an aura of disengagement from the work of the class. Such an attitude is fatal to any real learning, in effect becoming a self-fulfilling prophesy of educational failure.

Fortunately, such a resistance to learning is exceptional. Most younger managers who are on the rise are also in an up-beat mood about their own self-improvement, and they take advantage of every opportunity that comes their way. The older managers or executives, who are relatively more secure and relaxed about themselves and their careers are usually nevertheless ready to accept a program on its own terms. They are sufficiently confident that they can leave the job for a while when a good reason for doing so presents itself. And they are mature enough to realize that there is always more to learn, that nobody has the final answers.

Thus it appears that usually an executive education program is handed the advantage of an automatically preselected student group that is favorably disposed to a formal educational experience, provided that is well presented. If it is not well done the participants are quick to show their impatience. They have little tolerance for anything that they interpret as a waste of their time, such as irrelevant or trivial objectives or material; dull or vapid lecturing; cant in thought or expression; rudimentary, slow-paced, or obscure theorizing; or ideas pressed on them that oversimplify or run counter to reality. They are willing to argue about what is "real," but not for long. An instructor must earn credibility with them.

Once classroom rapport is established, all kinds of heresy is admissible even if not always accepted. Rigidity of mind is not a notable characteristic among executives in the classroom. They are engaged in a kind of life work in which a premium is placed on pragmatic and fast-paced problem-solving action. That is what they enjoy, and they carry that attitude directly into their classroom work, if given half a chance to do so.

That problem-solving orientation of the participants is accompanied (but to less obvious degree) by a little-noted but fascinating counterpoint: a hidden, almost subliminal, curiosity about the theoretical underpinnings of much that goes on in their outward world of affairs, or impinges on it from even broader reaches of human concern. Their life of demanding technical action, as measured by cold quantitative values, has so dominated their attention that by midlife many of them begin to feel its confinement. They are missing broader and more personal ranges of thought, of appreciation, of consciousness, of living. Consequently, they often see the executive program as an opportunity to catch their breath, so to speak, and to generate a perspective view of both their work and their personal direction in life, as some of them put it, "to recharge their batteries." That mood in an executive program fosters the desire to explore cause and effect and to derive principles, not simply to construct discrete answers. As is often said, there are no school solutions. Accepting that dictum constitutes a major intellectual shift for some. School was never like this!

As a class of people, executives tend to have active and inquiring minds. Every group will have some especially articulate members who, together with an instructor who has gained the confidence of the group, will stimulate everyone into exploring how things work, why things happen. An alert and prepared instructor will depart from his or her teaching notes and either assume the role of facilitator or deliver a minilecture. Achieving a momentum, the group is ready to examine new

data. Why are there different styles of leadership? What are the historic
origins of some of our more intrusive present day social phenomena?
Why is there a vibrant moral or ethical dimension to this or that situation? Why did that organization's profitability erode? Why did such-
and-such a corporate structure work in this instance and not in that?
Why didn't we spot that current assets were out of control? What effect
will this fiscal, monetary, or incomes policy have on our industry,
competition, or customers, and why? What are the major forces at work
in the international scene? What is being learned about the connections
between physical health, emotional well being, and personal productivity? And so on.

If the classroom learning can take place through the more familiar
milieu of active participation in exercises and discussion rather than
merely in listening and reading, then most executive students—being
bright, experienced, and adaptive people—soon pick up the rhythm in
the give-and-take of discussion and become good contributors in class.
More significantly, simply by expressing themselves they become their
own best teachers. And that becomes a prime ingredient in the first step
of learning: exposure to new ideas, particularly if they appear to be your
own.

When a program can achieve lively and creative discussion about the
abstract principles underlying concrete situations, then concern about
student attitude becomes largely immaterial. However, there are other
factors that affect student performance. Future attendees and the organizations that sponsor them should be aware of those influences so that
that the educational experience may have a better chance to live up to its
expectations.

Motivations and Apprehensions

We all know that there are wide differences in character, personality,
and ability among executives. Yet one of the most striking features of
executive students in groups is how much alike they are. University
professors who moonlight as instructors for in-company programs claim
that they can sense differences in group personality from different companies. But in front of all such groups these itinerant professors will use
the same or similar study materials, teaching notes, and instructional
methods; and they will attempt to establish much the same kind of

classroom atmosphere irrespective of group age, experience, seniority, or organizational source.

The effect of the classroom environment is to smooth out any tendency to deviate from the accepted norms of our executive subculture, at least during the cautious first few days of a longer program. Outwardly the participants reflect their predominantly middle class values, the common denominators in their prior educational experiences, the standard mores of their organizational life, and the generational myths and mannerisms that mold their concepts of how they ought to think and behave. Thus they have been already well trained, tested, conditioned, and screened before they ever reach the executive classroom. That tendency toward a general homogenization of the executive population allows us to identify some additional aspects of their motivations and apprehensions as participants in executive programs, especially in those programs that are two weeks or more in length.

The motivation of executive or junior managers to attend a program, and the frame of mind they bring to it, are influenced by the prevailing attitude of their organization toward the idea of continuing education, and also by the manner in which they are selected (or permitted) to attend. If time away from the job is granted grudgingly and the costs of doing so are resented, then the participant will be more likely to find the program a frustrating experience. On the other hand, if the organization expects its managers, whatever the level they may have attained, to take time off periodically for their own development, then even the busiest executive will make the time to concentrate on the course work, since that is the task that is mutually agreed upon for him or her. It is a part of the employer-employee social contract.

Most executive programs do not attempt to test or grade their participants; in fact program directors and faculty will not comment on student performance. Those few programs that tried to do so have since abandoned that feature. They discovered that combining assessment with this kind of learning was an incompatible mix of objectives. It resulted in tensions that blocked learning, and the measurements had to be artificial. The apparent exceptions are those special programs in which assessment is an avowed and preaccepted purpose; and even then a large share of the burden of assessment remains on the shoulders of the student. (In Chapter 12 we explain the issues of assessment in an executive classroom.)

The real measure of a person is his performance on the job. Responsibility has to rest squarely on each participant to grade himself on what he thinks he got out of the program. Only he can know. That fact underscores the hacknyed but true dictum that all development is self-development.

In a newly formed group, and after the barriers of initial strangeness have mostly been broken down, a surprising number of executives will admit that they approached the "opportunity," as it is euphemistically called, with a certain amount of trepidation. An example of that kind of anxiety took place some years ago. A nominee for a certain general management program happened to be a well-regarded manager of manufacturing. He had several hundred people in his organization and many years of service, having started out as a stock chaser on the factory floor. For several years he succeeded in avoiding the program, but finally did attend, and was a respected contributor to it. He later revealed that he had at first shied away because, having limited education, he would be "showed up" in front of his more erudite classmates. However, among other things, he discovered that book-learning does not necessarily confer common sense or an ability to work with people in the worldly affairs of business operations. And through his years of practical experience and habit of self-study he was at least as well "educated" as any in the classroom.

In a less dramatic way many executives harbor a slight degree of concern over competing so vulnerably with all those bright people in the class, until they discover that they too can pull their own weight, just as they do in the real world. A few others in self-defense will adopt a cynical attitude toward the course work, telling themselves that it is a distorted or oversimplified reflection of reality. But a skilled instructor can dispel that cynicism, as can also their more receptive classmates.

However, as stated before, most executives make good students and learners. If a few of them seem to withdraw or be turned off for any reason, they are soon brought back by the same objective, pragmatic, and problem-solving attitudes that earned them their positions in real life. They have already learned to accept the world and themselves with a kind of benign spirit. The word is: nothing is perfect, but we keep trying. If doing this will help, then let's give it a try.

Furthermore, if there is a good match in a program among students, faculty, subjects, and program objectives, then only the most obdurate of individuals could fail to experience moments of mental stimulus and discovery, and get pleasure out of the give-and take of discussions around many subjects. There will always be down-times, "sophomore slumps," but in a good program the group attitude always quickly bounces back.

The Private Agenda

Without intending to overstate the case, we may note that an occasional attendee enters a program for an extraneous reason. Her private agenda may have little to do with the educational and developmental purposes of the program. It may take different forms. She may be only complying with an established ritual of her organization. If she derives any value from the program, so much the better, but she will be slightly surprised if that indeed happens. Or she could see the program as being an acceptable gapfiller between a change in jobs. Or she regards it as a reward for long and faithful service, and therefore does not have any expectations to help guide her responses to the program. Or, in an occasional worst case, she may interpret attendance as a veiled signal that she is being parked on the shelf (that is, she can be spared). Her private agenda, then, is quietly to look for a new job, and the course work takes on a low priority in her mind, and becomes a cipher in her development.

Some people physically attend an executive program without being there in spirit, simply to get credit for having attended. They bring a small suitcase full of office work with them, and every moment out of class they are shuffling through their papers or talking on the long distance telephone. They may later recall the setting in which the program took place but will remember little else about it and will have gained nothing from it.

Then there is the unfortunate individual who shares a private agenda with his boss: he is there for a remedial purpose, an attempt to correct some alleged deficiency that is affecting his performance on the job. However, such actions are seldom effective since executive education programs are not designed and conducted for that purpose. Every one of us, as we move through our careers and take on new responsibilities, needs to strengthen our knowledge, skill, and perceptions in new ways, bringing ourselves up to date in developments beyond our work experience and preparing ourselves for the future. Such needs are a normal and motivating characteristic of personal growth and as such are addressed by executive education. But such on-going needs should be distinguished from some current and overt personal deficiency. For example, if an otherwise competent manager is universally known as abrasive, autocratic, and manipulative in his dealings with others, he might benefit more from some kind of sensitivity training than from the usual executive program, even though the latter may consider problems of that kind.

A substantial number of participants see the occasion of a longer program as an opportunity for working out a personal reassessment. Being removed for a time from the obligations of their work and family life, they can think about where they are and where they are going in their careers, and reconsider or simply reaffirm what they want to do with their lives. Although by all outward standards they are successful people with bright futures, they share in varying degrees those doubts and indefinable restless feelings that beset nearly all of us in the midstream of life. They have reached a level of maturity where they can be conscious of missed opportunities, of their strengths and limitations, of others who seem to be doing better, and especially of their own changing values and aspirations. And so they want to sort things out.

This private agenda of self-assessment and redirection doesn't frequently reflect a midlife crisis, as is so often overdramatized in our popular culture. It is more a normal result of quite healthy processes of maturation, a continuing search for authenticity as a person. Executive education literature will often express all this as being a prime value to be derived from the experience. Yet within that context of normalcy one can observe among executives some patterns of expectations, motivations, and apprehensions.

Aware and Willing Students

Most participants in well-established executive education programs are well-rounded individuals with no critical deficiencies in their competencies insofar as their current situation is concerned. During the program they work conscientiously to make the time pay off well for themselves and for their organizations. They show the same good qualities of conserving time and applying energy that led to their selection as executives and attendees in the first place. They know that the programs are expensive, that their time away from the job and family is disruptive, and that for the better programs competition to enter is intense.

Furthermore, since they are usually upwardly mobile and have already been tempered by a variety of work situations and responsibilities, they are more aware than are the drifters through life of how much more there is to know than they have yet learned. They can sense rather than articulate that need for broader and deeper understanding; the course work makes that need explicit for them. A fairly common example is in the field of management accounting and financial analysis. If they previously have had no training or experience in those subjects, they sometimes find it difficult but revealing to gain an overview of "the language of business," and they resolve henceforth to get better acquainted with their accountants. Even the financial people in the class who may have been immersed for years in some special aspect of the subject, appreciate an academic refresher and often pick up new ideas because the field keeps changing.

All of which leads us into Chapter 9 for a discussion of how program participants respond to specific subject areas.

9

Subject Areas in Executive Education Programs

> We should expect people to continue to learn and grow in and out of school, in every possible circumstance, and at every stage of their lives.
>
> JOHN GARDNER

Executives attending any kind of education program should expect to derive some permanent benefit from it, a due value received for their time, money, and psychological investment in it. However, any program is built up from individual modules that exert varying impacts on each of the students in the group. Although the program sponsor always strives to achieve for each student a total impact that is greater than the sum of its parts, nevertheless the effect varies from day to day, topic by topic, with each instructor, and within the ever-changing mood of the group.

If we are to be concerned about the quality of the educational experience, which is always a personal experience, then we have to consider in general how participants personally respond to individual subjects. That response in part becomes a function of the interaction between the type of individual student and the way in which he or she perceives the subject as it is presented. The different subjects, topics, and educational objectives that exist within the universe of executive programs, and the variety of ways in which they are treated, are so extensive and change so quickly that it would be impractical to list them all, much less describe them.

However, it is possible to comment on the usual responses of participants to some of the generic subject areas that occur most frequently in the programs, either singly or in combination. The comments hold whether the programs are university-sponsored, company-sponsored, multiweek in length or extending but a few days; for the generic nature of these broad areas lends itself to wide ranges of program treatment. (See Chapter 11 on external, or public, programs.) After all, among all those settings the students and instructors are more alike than different, having shared in a common wealth of experiences, ideas, literature, and aspirations.

Program Topics

Business Strategy

The study of strategy (sometimes called business policy) is often seen as the symbolic center of the longer executive programs that focus on the nature and issues of general management. The subject stands out sharply in the minds of most participants because it is often their first effort to study in a systematic and analytical way a business in its totality, and to form and publicly express their judgment about its past and present condition, future prospects, and where and how to take it over the long term. Making that effort marks a kind of milestone in the careers of the participants as they shift attention from specialist to generalist.

Studying strategy is a mind-stretching exercise in thinking of a business "in the round." It inspires the participants to want to understand more deeply about many related managerial matters, or at least to see their relationship to the whole. Thus many participants see business or organizational strategy, or the thinking processes that go into formulations and carrying out a strategy, as the binding force or the rationale for the kinds of subsidiary operating decisions and actions with which they are more familiar in their current operation responsibilities. On the other hand, those participants whose responsibilities already include long-range business planning or leadership of a business, find that the strategy discussions strengthen their own level of competence and comfort in their job. We have yet to discover an executive who thinks he or she already "knows it all," after having participated in one of those challenging give-and-take sessions with a group of other smart and experienced people. There is always something more to learn, even for the instructors.

However, some participants, or the superiors who have some control over their careers, find it hard to distinguish between the processes of strategic planning and the mental capacity to think and act with strategic logic and vision. The processes produce paper plans; the thinking produces business performance and is an essential ingredient of leadership. Yet the processes are needed to support the thinking with substance, communication, control, and continuity. General management programs will try not only to instill a sense of what it means for senior officers and their staff and line leaders to generate a common strategy, but will also explore the ways in which organizations and their managerial regimes succeed or fail over the long term. For instance, if the management of the budget, of the human resources, of purchasing, of the sales force, of market forecasts, of production cycles, of the pace and direction of critical technology, of the actions of competitors, if any of those and other significant factors are inconsistent with the strategy, then something has to give. Thus it is that strategy sessions are not just head-in-the-clouds dreams, but rather exercises in learning to act according to a best estimate of the situation, and knowing when and how to change over time.

The strategy sessions in an executive program address the foregoing concepts. Yet within any group of participants there is some diversity as to what the individuals get out of the sessions. Certainly a good case discussion can leave its mark on nearly all the group. But if the managerial climate and the perceived reward system in a participant's home organization is focused entirely on short-term operating results, the study of strategy is likely for that participant to be a waste of time and money. However, most progressive organizations understand that their survival and growth depend on how astute and creative they are in facing the future of today's complex world; that is, they understand that their fortune depends on how well they formulate and carry out their business strategy while still meeting their on-going commitments. Thus the strategy discussions become the centerpiece of most long programs, even those whose perspective may be primarily from one of the functions.

The External Business Environment

Any time the leaders of an organization seriously try to formulate and carry out a long-term operating strategy, they become crucially aware that every aspect—economic, social, physical, technological—of the perpetually changing external environment is important to them. They learn the necessity of having their antennae out, scanning many frequencies throughout the local, national, and international environments, and reading those often confusing signals correctly in their prognoses of the future. For the future is what strategy, leadership, and personal development are all about. And the future resides in what is about to happen everywhere, all of the time, while mutually affecting every element of the world.

Achieving that kind of awareness is a large order. Yet it can be well argued that the ill fortune of many an enterprise can be traced to the failure of a managerial regime, often a previous one, to read properly the significance of changes taking place in those external environments. The point does not need belaboring; we see all around us radical changes taking place, perhaps most dramatically in the international arena. Those who interpret those changes most effectively learn how to think about the forces of change, and therefore sense the cues before others do and take the initiative to act on them.

But there must be a frame of reference for thinking constructively about those grand matters. To provide such a basis, executive education programs are paying increased attention to external environmental topics. These ideas are brought out in case discussions and in lecture-type sessions led by professors and other leading thinkers in sociology, economics, history, philosophy, international trade and finance, and ethics; and in a myriad of other topics, issues, and intellectual approaches. For the most part such sessions (and, in some instances, whole programs devoted to external environmental topics) are well received by executives; their experience has already convinced them that these are important matters.

There are, however, some inherent difficulties with this subject area. Many executives have only a narrowly prescribed background in their education and experience, and, indeed, in their range of active interest. They have not thought much about what goes on outside their professional niche; they are "all business" and proud of it. As a consequence, valuable podium time is spent simply bringing the group up to speed on basic facts and concepts. In the relatively short executive programs, then, there is little time to develop any subject in depth. Therefore the only hope is to stimulate the participant's interest and desire to pursue the subject further. Finally, there are those who believe that the topics are too "academic," or who think that they are self-evident. That category of participant is sometimes surprised to discover in the classroom that for the first time he or she translates passive awareness into proactive business insight.

Fortunately, alert executives have already been and continue to be engaged in the best of all liberal educations, which is an active and productive life. And so the task of the classroom sessions on the external environment of business is mostly one of providing new perspectives on matters that the student already "knows" but realizes that the need to understand better. That goal makes for a different classroom atmosphere and student–instructor relationship than is ordinarily found in the undergraduate classroom or lecture hall. Many a heated argument, a clash between different backgrounds of experience and well-founded viewpoints, ends up as a stimulating learning experience for everybody. That experience is group learning at its best.

The Internal Business Environment

In general, topics pertaining to the internal business environment, which appear in one form or another in nearly all executive programs, cover the internal operating processes and relationships that define an organization, that is, the "machinery" of the enterprise or its inner workings. One approach to those inner workings was made famous by Douglas McGregor (1960) in his book, *The Human Side of Enterprise*. McGregor emphasizes the concepts of individual psychology, shading into sociological issues as the groups under examination become larger and more complex, and finally culminating in the macrovisions of business as a human activity. That activity influences or even creates its own external environment, with all the economic, social, and value-laden implications of that phenomenon for the nation and the world. (From that perspective, our distinction between internal and external environments is only an arbitrary device that helps us to sense relationships among different viewpoints on matters of managerial interest.)

In their most elementary forms, human-relations sessions are training sessions in "behavior" (to use the word in the jargon of psychology). The participants are drilled in a panoply of insights about interpersonal relations, leadership, communications, problem-solving versus frustrated behavior, group dynamics, motivation, life planning, and many variations on those themes. Although they are played out in familiar business situations, the lessons apply to human beings in any circumstance.

The presentation of behavioral topics needs to show some sensitivity for the level of experience and the self-perceptions of the student group. Junior executives can respond well to the teaching methods that are widely used in college courses, whereas a group of senior people would likely find them obvious, trivial, or embarrassing. There are many situational cases in which the scenarios are familiar to them in their current positions, and such cases should be used at least for openers.

This dictum about sensitivity applies to any subject, but especially to behavioral topics. Perhaps the tough and technically oriented self-image that is fashionable in our culture makes it unusual for us to try to understand why people behave as they do. Or maybe some topics strike uncomfortably close to home, and defense mechanisms spring up. In any event, the sessions elicit strong reactions, and their value appears to be heavily instructor-dependent. Given the right conditions, these sessions will draw top honors from many of the students, or so they say. We have to sort out their human appeal from their long-term effectiveness, and that effectiveness remains a controversial point.

Another observation worth noting is that the human side of enterprise is not separable from the "technical side" as they work together in real-life operations. Ill-conceived organizational structure, policies, or operating systems and practices can lead to "people problems," and vice versa. It is the continuing and ever-changing interaction between those two sides that advances or impedes achievement of organizational goals. Because of that interaction, most executive programs in general management stress using so-called rich cases, actual situations where cause-and-effect or even "the problem" is not readily apparent but is buried in the messiness that characterizes our lives and work. If inventories are out of control, or product quality has deteriorated, or sales have taken a nose dive, those situations are often evidence of trouble within the many-faceted network of factors operating in the underlying business reality.

As a consequence of that complexity of issues within a business, executive programs put considerable attention not only on the human side but also on the technical side of the internal environment. Of course the degree to which they do that depends on the program objectives. For instance, a session, a series of sessions, or even an entire program may focus on a selected primary function such as marketing, engineering, manufacturing, and so on. Or it may confine itself to one of the sub-functions such as quality control (or quality assurance, as many call it). However, if the sessions concentrate only on the operating-manual techniques of the quality control function, then they are really training sessions that should be attended mostly by specialists in the activity.

If the leaders of an enterprise see as a key objective of the organization its attaining and maintaining improved levels of quality throughout all their products, then they are likely to discover that nearly every function, human and technical, has some part to play in attaining that objective, and that the human side is intimately bound up in both the problem and its solution. That fact calls for concerted educational effort by many executives throughout the enterprise; first learning together, and then continuing to act together based on that learning.

The discussions that take place in both in-company and public executive programs, and the way the subjects are set out, direct the attention of the participants beyond details and into the broader areas of cause and effect, into the higher levels of issues and objectives, of means and ends. A program that purports to be concerned solely with the marketing function soon discovers, as it probes beneath the surface of "marketing" issues, that other matters are also involved. So, the marketing program as it evolves over time takes on the heterogeneous nature of the generalist program. We can see an analogy in the child who paints with raw colors straight out of the tube, whereas the mature artist mixes colors, either physically or in his mind's eye, knowing that nature does also. In business, as in painting, that mixing of different inputs to achieve certain outputs might be called suprafunctional. We see it taking place in sessions dealing with interfunctional coordination; with strategy formulation; and with planning and control systems. But the blending of diverse business considerations into a coherent picture must be done with skill, or, like the inexpert painter, you end up with mud. The need for expert blending occurs especially in the subjects of accounting and finance, subjects of increasing importance in many executive programs.

Business Economics

Accounting and finance are sometimes called the language of business. It is the language by which the concepts of business as an economic activity are derived, expressed, and applied to the planning and direction of operations. A manager who does not think in those terms can hardly be called a businessperson regardless of how strong he or she may be in other aspects of administration.

The successful small businessman quickly learns to define his business in terms of the numbers it generates and will project into the future. He is forever managing both the income statement and the balance sheet. He thinks about the business as a microeconomic unit. His managerial decisions are fundamentally economic decisions, for he knows that it is along that dimension of measurement that his business will survive or fail.

In contrast to the small businessman, the manager or junior executive inside a large or complex organization lives in a world where there is not only a division of work but also a division of knowledge. The kind of working knowledge that he or she finds herself most often separated from is financial. Through the lessons of experience the young manager in sales or engineering, for instance, absorbs some notions about the financial condition of the business, and may even learn to spout a bit of the lingo. But her understanding is shallow and her interest remote. She sees that subject as being the esoteric job of the financial fraternity. If she does her own job well, she thinks, someone else will see to it that the numbers come out all right.

As the specialist-manager begins to rise up through the executive ranks of a complex organization, she realizes that her decisions and actions can no longer be based solely on prior technical knowledge and skills. She starts to be held accountable for the business consequences of her recommendations and actions. Therefore she needs to strengthen her own working knowledge of the microeconomic factors that should help to shape her decisions. She must be able to explain her choices among alternative economic courses of action, and do so convincingly.

Of equal importance in her growing inventory of skills, the generalist-manager must know how to enlist the active participation of trained professionals during the planning phases of her future courses of action. Too often the professionals in accounting and finance are informed of decisions after the fact, thus finding themselves relegated to the role of recording the past rather than influencing the future. (We might interject that this is also true of human resources managers; but in both cases the situation seems to be improving.)

Most of those managers who think seriously about the total picture of business they are in and about their own future in it, recognize the foregoing needs. But a large number of them also recognize the inadequacy of their understanding, and so they both welcome and fear the executive education sessions in accounting and finance; welcome them for strengthening their comprehension, fear them for exposing their ignorance. Fortunately they quickly discover that many of their classmates feel as they do, and so they settle down and unabashedly "push the numbers" in a conscientious effort to understand the principles underlying them.

Because of the equivocal mood in which executive education participants often approach sessions in management accounting and finance, and because of their practical experience in other aspects of managing, the sessions are taught in special ways. The intention is not to make experts out of the students, but rather to help them to learn how to look behind the numbers, to use those numbers as only one more means for diagnosing the health of the business and identifying and evaluating possible decisions about it. Bookkeeping is essentially ignored, and only passing references made to so-called accounting conventions. Furthermore, just as in the case of sessions about the human side of enterprise, the specifics of the financial subject matter are made relevant to the experiential levels of the participants, and appeal to them as representing real situations. Instructors who can meet these special challenges are apt to foster lively sessions and are much sought after in the world of executive education.

Although the sessions in accounting and finance cannot totally dispel the aura of mysticism that seems to surround the subject, at least they can give the nonfinancial manager the motivation and courage to consult his financial manager or accountant during decision making. We are reminded of a story from some years ago. Senator Vandenberg was being petitioned by the Pentagon to have Congress bail it out of a disastrous program, one that the Senator had little to do with. His reply was, "If you want me in on the crash landing, you'd better have me in on the takeoff." It is rewarding for an executive program director to hear a participant say that from here on he is going to have his accountant included in his "take-offs."

Modes of Thinking

Executive students are more likely to respond with interest and, therefore, with benefit to themselves, if they see a subject as being relevant to personal needs and to the kind of organizational positions they occupy. If the subject or the way in which it is presented calls for the kind of thinking which they are comfortable with, so much the better. For those reasons, we can better understand the differences in student response to different subjects if we consider the dominant kinds of thinking skills that may be represented among the populations attending executive programs.

An observer of participants in many different kinds of executive programs can discover consistent patterns between how executives think in the classroom and how they act in real life. The classroom becomes a kind of controlled laboratory experiment, in which the thinking (or merely reacting) processes reveal themselves more clearly than they can above all the noise and turmoil in real-life situations. If the observer has a broad operational background, he or she can sense connecting relationships between modes of thinking and the demands of organizational action, that is, the various ways in which different forms of intellectual input support different forms of operational output. The wide range of subject matter in executive programs, combined with the wide range of executive participants who work on those subjects with relatively uninhibited intellectual freedom, weaves a rich fabric for displaying the varying operational significance of different modes of thinking.

In Chapter 7, we present several ways in which kinds of work or output change as a person moves upward through a managerial career. Those categories were objective; that is, at a specific organizational level, a certain kind of work, problem, or responsibility predominates. But the question remains: What different kinds of thinking do such changes call for? To answer that question, consider subjective categories; the mental caste of individuals that should match the different kinds of work to be done. Previously we considered how executives responded to external stimuli, to the demands of their assignments. Now we are looking at how they respond to internal stimuli, to their own unique mental processes and innate interests. Table 9.1 helps us to visualize that relationship.

Note that (1) the table includes nomenclature that differs from any used previously. It is describing a different, even if related, phenomenon. For instance, we did not previously use the term *leader* as a distinct organizational category, nor do we mean to do so now, but rather to use it as a value-descriptor of certain persons at any level of an organization who are seen as more than competent technicians or ad-

TABLE 9.1   Modes of Thinking

	Specialist	Manager	Executive	Leader
Analysis	xxxx	xxx	xx	x
Synthesis	xxx	xxxx	xxx	xx
Evaluation	xx	xxx	xxxx	xxx
Integration	x	xx	xxx	xxxx

ministrators. (2) We do not intend to imply that one way of thinking is superior to another. They are simply different, but not mutually exclusive. (3) This discussion does not address a possible third dimension of differences, that of social attitudes and values. For the most part the participants exhibit, either because of birth or absorption, the usual middle-class values that appear to be increasingly shared among all the industrial nations. Presumably the participants, like the national population in general, tend to become more conservative or parochial with age and experience. Although we must leave the topic of social attitudes and values to the sociologists, its importance should not be underrated. Any well-rounded general management program addresses sociological, cultural, and value systems, and ethical issues affecting mankind, from the point of view of the individual on through organizations to society as a whole. Many participants find these topics to be the most stimulating portions of a general program; and as shown in Chapter 11, they constitute the entirety of some well-known public programs.

The abstract relationships shown in Table 9.1 have a practical bearing on the task of matching a subject to the audience. For example, in the field of accounting and finance, for junior executives, it would be more appropriate to bias the subject material toward the managerial accounting end of the scale, while for senior executives the bias should lean toward topics affecting the corporate financial structure. As another example, in the area of human relations, lower level people would benefit most from emphasis on universal behavioral topics, whereas senior people might find more interest, for instance, in examining the many facets of organizational politics. The subjects are the same, only their scopes differ. And in a subtle way the kind of thinking changes from step to step up the career ladder.

The dominant mode of the specialist is analytical; his task is to discover and organize facts in logical and objective ways and to draw conclusions therefrom. Depending on the degree of sophistication of his specialty, he may also synthesize facts into higher order findings about their operational significance. The "high specialist" may carry through to the application of value judgments, in which evaluative and integrative talents become essential not only to problem solution but to their identification in the first place. Examples of high specialists include the scientist in the corporate laboratory or the operations research specialist on the corporate staff.

The new manager, that person who has taken a giant first step into a new world of responsibility, finds that she must now synthesize nonmeasurable sets of data, such as reconciling the technical demands of her organization's work with the psychological terrain of her subordinates, with competing or conflicting needs of her peers, and with the often unclear or ambiguous objectives and predilections of her superiors and of the organization.

The established executive, on the other hand, usually finds that his main effort is to apply value judgments to decisions about opportunities, problems, issues, or corporate intentions in the face of uncertainties. He is being judged by his superiors not only on how his organization currently performs but also, and perhaps more important, on the quality of his judgment over the long term.

Finally, the organization's leader provides the element of integration that defines the organization and places it legitimately in its environment. He provides to it those qualities of unity, emphasis, coherence, and vision that can come from no other source, and that are essential to identifying and achieving organizational purpose.

By recognizing the fact that emphases on different modes of thinking will shift as the manager progresses up a career ladder, it can be concluded that the character and content of education programs also should shift as the individual experiences them through the increasing years of his or her career. But the shift should be in perspective; the underlying truths of the subjects endure.

Educational Progression

If experience builds on education and continuous learning, the reverse is also true: learning builds on experience. The on-going interdependence of experience and education indicates that an individual should actively engage in an array of educational experiences throughout his or her lifetime. The activity should be neither haphazard nor repetitive, but instead selective, focusing on a planned and personal development schedule that is gaited to changes in career needs, aspirations, and prospects. Education and experience will work together in support of continuing high-quality performance on the job, which itself is the final source of competence and personal growth.

In America the popular view is that in the pattern of a lifetime, education ends and then work begins, continuing into idle retirement, and culminating in old age and death. However, in many professions such as law, medicine, science and engineering, the church, and the military, work and study commingle throughout a career. So much keeps changing that there is always more to learn. If not tested against new realities, old ideas lead to rigidity of mind, and are quickly outflanked by the stream of events. The mind must be exercised or it will atrophy.

Likewise, a static business education program is a dead program. Ten years ago we scarcely knew of the lessons to be learned from the Japanese. Today we listen carefully; tomorrow it will be something different. For individuals, a single general management program should not be the end; there is no end. The Young Presidents Organization conducts exciting and well attended seminars for its select membership. Comparable sessions are conducted by some universities and other institutions exclusively for chief executive officers or their surrogates. So we see that in speaking of executive education we speak not of a program but of a lifelong commitment, one in which no subject is ever finally learned, but is forever being relearned in new ways.

10

Educational Policy and Administrative Decisions

> Institutions as well as individuals are the clients of adult education. And through both the institution and the individual, society is the client.
>
> MALCOLM KNOWLES

When an organization decides to use executive education programs as one of its actions for the continuing renewal of its executive resources, it will find itself having to make a number of policy and administrative decisions. For example, if the organization is fairly large, it will be faced with the question of whether to have its people attend university sponsored programs, or to do its own programs, or to use a mix of both. But first, and more important than any such operational policy, the organization must be clear in its own mind about its attitude toward continuing executive education in general.

Corporate Attitude

As in anything else that the organization sets out to do, the attitude of the corporate leadership will largely determine the success of the effort. In the case of executive education the leadership must genuinely believe that well-conducted continuing educational effort is intrinsically worthwhile for its managers and for the company. The sincerity of that belief cannot be faked for long. It is too easy for the organization to spend time and money, and then at the first drop-off of enthusiasm or the first sense of disillusionment to allow the effort to wither away. However, as

93

in the case of product research and development, the results of continuing education can show up only after sustained effort over a long time.

The public relations value of a development program, whether for a new program or for the people of the organization, can easily mask its true value. It is tempting to undertake the forms of education without the substance of learning simply because the leadership thinks it is a "good idea." It will be a better idea if their faith is based an understanding of the developmental and educational processes as described in previous chapters. But in addition to those somewhat abstract matters about attitude, there are some operational considerations to take into account.

Make or Buy?

That familiar question clearly applies to the function of executive education. The obvious answer for a small organization is that it has no practical choice. It cannot afford to make, therefore it buys the educational capability by having its managerial people attend some of the many excellent programs offered by colleges, universities, associations, and other professionally based institutions selling those services in the public domain (see Chapter 11).

Once an organization reaches a sufficient size in executive population and financial assets for sustaining the educational effort, then it has several options open to it. It can continue to buy, or start designing and conducting its own programs, or do both. There is no one best way for everybody. Much depends on the character of the organization, on its felt needs, and on its historical precedents. Most companies turn to public programs as their primary resource, but supplement them with an occasional internal seminar or company retreat.

Some large organizations have long histories of supporting internal education and training centers as well as using outside programs. International Business Machines (IBM) is an example of that practice. In contrast, General Electric depends almost entirely on its own internal programs for executive education. Each of those companies is well known for the development of its executive resources. Neither system is inherently better than the other. The point is that whatever practices any organization follows, they should fit naturally into the kind of organization it is, responding to its own special needs and to the expectations of its people.

If the organization strains at some educational preconception of what "ought to be," it can be headed for disappointments, as many a company (and university) has found out. However, once it understands what executive education can and cannot do and how the process works, then it should be guided by a balance among actual needs, available resources, acceptable quality of outcome, and inherent risks, just as for any other investment decision. But the basic consideration should always be for a running appraisal of educational needs, which are ever-changing variables in the equation.

Defining Needs

Organizations adopt a variety of means to help them define their needs for executive education. The processes for doing that sometimes appear under labels such as "needs analysis." Depending on how the organization customarily works, the processes run the gamut from formal reports, task-force studies, oversight reviews, design teams, and the like, to informal surveys, discussions, or individual staff initiatives. Large companies with complex structures and diverse needs may use any of those processes to fit the occasion. One of the most effective ways to find out what is needed is simply to experiment with either a made or bought program and see what happens. It may test to destruct, but it will also instruct. As in most things, one learns by doing; theory comes later.

Whatever researching processes the organization uses to help it determine its educational course of action, the people who do the legwork must tune their mental screen to different frequencies and integrate a set of images along different dimensions. Previous chapters discuss the clues to identifying needs, but they are mentioned here to put them in the context of our present purpose.

1. A strategic view of the company: Where does its leadership think it is taking the company? Why, when, and how? The executive education staff is not likely to get clear and specific answers. It has to pick up clues and translate their implications into educational needs. The company leadership cannot make that translation; that is a function of the education staff, which must do a bit of sleuthing.

2. The present and probable future demographics of the executive population: What are its stratification layers, numbers, movements from one layer to another, patterns of job requirements, patterns of experience, and so on? How are those factors changing over time? The implicit assumption is that the continuing development of the executive resource should be just as perpetual a function as is the existence of the organization itself.

3. The intellectual and psychological patterns of the people within the managerial populations: How do they work, think, learn, and change in their aspirations and competencies as their jobs, responsibilities, accomplishments, and ages change? The best data comes from the people themselves, by the researcher or program director talking with and observing them both on their jobs and during a formal program. That takes time and effort, but the pay-off is added realism in the educational program, a valuable and sometimes scarce ingredient in some educational endeavors.

4. The political terrain of the organization: How do the attitudes toward executive education vary throughout the organization? Where are the advocates, the skeptics, the nay-sayers? Education is an overhead cost item, therefore its continuing support can be precarious in spite of declared corporate faith in it. It needs money, and so it needs selling, careful and well-paced presentation, and the building of a good track record and of a supportive clientele.

Internal Programs

Motivation for Internal Programs

In the preceding discussions of internal programs that there are always two customers: the individual attendee *and* his or her parent organization. Each benefits because each depends on the other and in a sense shares the same fate. To that extent their needs are intertwined.

From the organization's viewpoint, its motivations for doing its own programs may take one or several of the following forms:

1. The organization may be of sufficient size, and therefore have a sufficient upward flow of people though its executive ranks, to make it more practical and, perhaps, less costly to conduct its own programs. Thus it may more readily maintain the desired "critical mass" effect of repeatable programs throughout its executive populations. Under those circumstances it may sponsor one or several continuing series of generalist programs that are similar to those sponsored by universities.

2. The organization may see its internal programs as fostering organizational identity and cohesion, particularly if the company is decentralized and serves a number of markets with different technologies. That is a legitimate expectation; people studying together and working on common tasks in an educational program develop rapport, friendships, and even business relationships. Proprietary company matters can be discussed. The curricula, even in so-called generalist programs, can be tailored to emphasize company-specific interests, and each part of the company learns from the others.

3. The organization may have an extraordinary one-time need of an operational character that it can best satisfy through a custom-designed and saturation-effort educational approach. There are innumerable possibilities. For instance, the company may want to launch a campaign to revitalize the companywide strategic planning and budgeting processes; get control of its current assets; strengthen ethical awareness among its people; adjust to a reorganization or a major acquisition; instill a new marketing attitude; develop a management team for a complex new project; face into a new technological revolution affecting most of its products and markets; or engage in foreign markets and joint ventures. A program with its group of select participants offers a natural occasion for generating solutions to company problems and enthusiasm in pursuing them, with an improved understanding of what is required to be successful. Some companies may be unaware that the participative type of executive educational mode for engaging the company people in helping to solve a company problem has been proven to be possible and to work better than by only issuing company directives, jawboning, or pounding the drum at company revival meetings. The change is created through the people, not by fiat. They own the change, the ideas come from them, and they'll make it work.

A word of caution is in order. An organization that designs and conducts its own programs must be on guard against their becoming doctrinaire, myopic, or pedestrian: doctrinaire, by simply expounding entrenched company policy and practice; myopic, by not challenging the participants with unfamiliar and perhaps heretical concepts; and pedestrian, by not reaching beyond rudimentary ideas indifferently taught. Public programs find ways to avoid those pitfalls, otherwise they would run out of customers and die. Internal programs can avoid them by sharpening program objectives to specific audiences; by carefully selecting faculty for both design and teaching; by common sense in administration; and, above all, by having the willingness and imagination to experiment.

Faculty

Of all the factors that determine the quality and effectiveness of an internal program, the most influential is the group of resource persons: the teaching faculty, experts, practitioners, thought leaders, catalysts, program "enrichers." Two programs that may look alike in written content can be entirely different in effect because of differences in teaching resources and in how they are used. A good program design cannot substitute for important ideas well taught. "The program" is what occurs in the privacy of individual minds (see the section on program effectiveness in Chapter 8).

The heart of most internal corporate executive programs is designed and conducted by members of college and university faculties. The program director, who is the corporate representative, may determine objectives and lay out a preliminary design. But the final structure is largely a by-product of the complex processes that must be pursued together with university people in recruiting resources, discussing approaches, juggling schedules, and fitting together the pieces of program to make an educational whole. Each time a program is staged, it is a customized creation by all the resource persons who appear on it. Good programs cannot be Chinese copies. Each session reflects the independence of mind that is a hallmark of academia, and takes weeks of preliminary planning by all concerned.

There are several advantages in using outsiders in company programs. They are professional teachers, bringing with them all the potential for quality education that the term implies. They bring fresh and stimulating ideas from the experiences of other organizations and the world, which helps to combat the tendency toward company colloquialism. They are expendable; that is, tenure is not an issue, and that allows for an easy and healthy relationship. Although the best known are not always readily available and are likely to be expensive, many new faces keep appearing with fresh ideas and excellent teaching abilities. They are cooperative and enjoyable to work with. A program manager can build up over time a sizable stable of faculty candidates, thus affording production flexibility. Finally, they are a source of future expert consultation, which can be advantageous to both sides.

There are also some difficulties in assigning a major design and teaching role to the professors. Possibly the most obvious is that a few highly placed company people may, on general principles, take a skeptical view of what they see as an "academic" program. It is often impossible to get the instructors you want when you want them. Their commitments must be made months ahead of time (although a last minute referral will often save the day). They prefer repeat business, which makes it difficult for a small company to attract them. Furthermore, the professors want to build their reputations with well-known organizations, and thus open up the possibilities for consulting work and case writing. They are not cheap; their honoraria and travel and living expenses can easily be the largest direct cost item in the program's budget. Finally, there is the matter of safeguarding company proprietary information. However, experience shows that to be seldom, if ever, a problem. Outside teaching is an essential part of their professional work, and they are not going to jeopardize either their ethical standards or their reputation by revealing proprietary information inadvertently gained in confidence.

Any internal program is strengthened if the professional teaching resources are supplemented by management practitioners, specialized experts, or others who can apply working knowledge or experience to enrich the program and add balance to what otherwise might be perceived as too theoretical or abstract a presentation. The executive, the person of action, when engaged in a theoretical discussion, is haunted by the unspoken question, "so what?" However, the veteran of the firing line can bring reinforcing, clarifying, or contradicting insights into play, and thus intensify the learning process so that the executive participant can see how to answer that question for him- or herself. It turns out that organizations discover natural teachers within their own ranks simply by affording program participants the opportunity to share their knowledge, skills, and insights with others. When that happens, everybody, including the organization, is enriched.

A special case, and an important one, in sharing viewpoints between an adjunct resource person and a group of executive participants, is the session in which a chief executive, or surrogate, meets with the group at some point in the program. They exchange not only viewpoints, but also benefits. The group gains firsthand understanding of what currently concerns the top level of the company, and they read the signal that they and their development are important. The CEO gains a special view of the organization, one that is not likely to be seen through formal reports, staff meetings, or plant visitations. But gaining such an advantageous viewpoint takes a bit of doing or, shall we say, manipulation. The CEO must learn not to give a formal speech, but instead to encourage informal questions and discussion; and never to cut into happy hour! If the group has been together long enough to become at ease with itself, then the most stiff-necked of CEOs can also learn how to get the most out of a session of this sort. Even a CEO is not exempt from the possibility of learning something new.

Administration

PROGRAM DIRECTORS

You can see in the previous few chapters that executive education is primarily oriented toward developing insights about business organizations, as contrasted to training in individually applied skills. At least, that is a tendency that differentiates it from what you normally thinks of as "training."

Because of that strong operational orientation of executive education, it is important that the people in charge of such programs have solid backgrounds of experience in the management of business, major projects, or key operating functions. Their main function is not in teaching but rather in bringing the best and the brightest of academia and professional leadership together with groups of executives in a joint learning experience. That brokerage function calls for exercising good judgment that comes from personal knowledge of how people think and work in organizations; it calls for the instinct of the practitioner. To that qualification the program directors should add an affinity for the values and methods of the academic environment even though they may not have formal credentials in it. Armed with those special capacities, a program director or manager should be given free rein at the professional level to conceive, form, and conduct a program. There are critical educational judgments only the director can make, if true learning is to be achieved.

Thus we find that many a program director or manager is a well regarded midcareer manager on a rotational assignment; or, even more often, he or she is a successful and mature individual who has made a permanent career shift from operations to the field of executive education. Those managers who have done so discover it to be an exciting and rewarding field of work in the later portion of their careers.

CONTENT ORIENTATION

In the opinion of many participants, every program should be shortened by 20 percent, but to adopt their creative suggestions for other improvements, it would have to double in length.

That paradox is only one of the factors that makes life interesting for the program director. He or she is perpetually working for a balance—or perhaps compromise is the more exact word—among a number of variables such as enough time for a topic to do it justice without short-changing others; homework sufficient for reasonable preparation but without overload, irrelevance, or redundancy; informational versus participative sessions; and class size versus financial viability. And perhaps the biggest balancing act of all is covering enough solid content within the given time constraints to make the program a meaningful experience and meet its perceived objectives, yet not attempting so much as to make it superficial. We doubt that a detailed instructional manual could help much to solve those dilemmas. Insight to dealing with them can be developed by experience and can help to reduce their severity. However, a few general observations may be helpful.

A program should have a perceptible *theme* or *structure*. These can help to give the participants a sense of direction and progress, a feeling that the program is not just a hodgepodge of random sessions. For instance, a short program on managing current assets could be seen as a lesson on interfunctional coordination, or it could be treated primarily from a financial viewpoint. A general program might seek to build reinforcement among small-group business exercises and the specialized topics that pertain to them. Whatever the case may be, the program manager should know in specific terms what he or she is trying to accomplish in order to judge the proper length of the program, the role of each session, and the relevance of the whole program to the needs of the particular attendees for which it is designed. Generalized statements of intention do not help very much.

If a program can be successfully repeated it is probably a winner. Only a *repeatable* program will deliver a significant impact on the corporation. But no good program is exactly like its predecessors, nor should it be. It is subject to voluntary and involuntary changes, and constantly being adapted to corrections of weaknesses, to newly discovered approaches, and to administrative exigencies.

The executive student quickly senses whether he or she is attending a lively *intellectual challenge* or is merely being dragged through some ossified Chinese copy of ancient academic origin. Group disaffection is infectious and cumulative. Even if in such circumstance the successive groups dutifully go through the motions of being instructed, it is better to change the program leadership or drop the program altogether. A mediocre program is too costly in executive time and corporate funds to be tolerated. Perhaps an even greater cost is the disillusionment of the participants themselves with the processes of life-long education. In their disillusionment with a disappointing experience they will become educational drop-outs, and never know what they might have learned at those times in their careers when it would have done them the most good.

Fortunately, good spirits within a group are equally infectious, and as *unpredictable*. You can predict that there will be some difficulties with the beginning and ending sessions of longer programs, but you cannot predict if or when a "sophomore slump" will occur. With experience, a program director can usually predict the likelihood that an excellent instructor will do as well with the next group as with the previous ones; but the effect of a particular combination of sessions or faculty personalities cannot be predicted. Only experience will show what works.

The *size* of groups, the amount of *preparation* required of participants, and the average *level of operating responsibility* of the group, are three variables that influence the way in which a program functions; and they help to guide the program director in how it should be structured. As in any kind of advanced education, a session that consists primarily of a lecture can accommodate a relatively large number of participants; and a professional lecturer who is skilled at interacting with executive participants can interweave good discussions throughout the more formal material and keep a large group of people alert and interested. In contrast, some subjects can best be assimilated through study or active preparation ahead of time, and so the group is broken up into small discussion or working teams in which participation is encouraged through group pressure. A special case of that technique is the use of computer-assisted business simulation competitions. If sufficiently sophisticated in construction, they can become the high points of enthusiasm in a program.

As for individual background reading (in contrast to case preparation), executive programs of limited duration can tolerate only a minimal amount, just enough to sample some of the important literature. Most participants are willing to taste but not to eat, and a perceived overload of reading work quickly becomes counterproductive, just as does too much free time. Most participants expect to work and learn, not to have a vacation. But the rapid succession of new subjects and personalities becomes confusing and tiring; and so a good program director learns to pace the intensity of the program with care.

The program director should keep in mind the fact that older and more experienced participants will respond better to subjects when the material corresponds to their level or responsibility in their organizations. Clever teaching methods or games are usually not well received by senior executives. They prefer lessons to be learned from real-life cases involving organizations comparable to their own in scope and complexity. Furthermore, participants at any age appreciate variety in subject matter and teaching methods. The ogre of boredom is always just around the corner, and variety can help the group to avoid it.

OPERATIONAL BASIS

Most companies choose to operate their educational programs as one of the standard functions of their human resources staff. Some companies, rather than going it alone, join with other companies in a consortium to support a facility, faculty, staff, and a set of time-shared programs. That cooperation has been more so a practice in Europe than in America.

If a corporation is large enough to serve a number of distinct markets and is administratively decentralized, then the internal executive activities may be sufficiently extensive and varied enough to warrant a semiautonomous and central operation, one having its own campus and supporting itself, at least in part, though sales to its many divisional customers. That entrepreneurial relationship imposes a beneficial marketplace discipline on the activity, and in the process provides a fair measure of whether or not it is doing the right things and doing things right.

As you can imagine, advantages accrue to an educational operation that can support a "product line," a rolling inventory of many different offerings that respond differentially and quickly to the changing business needs of the organization and of its various executive populations. If they have a number of different programs underway, an in-place staff of program directors or managers can promote a synergy among their ideas and resources; the activity can better track the needs of the corporation and of its divisions; and the educational activity, like the centipede, is less likely to stumble financially.

The real estate facility in which a program is conducted should provide a pleasant atmosphere, but also shield the attendees from distractions. The work of the program requires involvement, concentration, and a good deal of group interaction. Some participants need to be reminded to leave their office work behind. For those reasons companies often try to select a facility that is not in a city, adjacent to its headquarters, or at a major plant location, and yet has convenient transportation services. Some public conference centers serve the purpose well. Large corporations often have their own meeting places, even if a large portion of their education and training work is conducted elsewhere.

Conclusion

It is obvious that for most companies the costs of fixed investment and effort required to conduct their own programs as a continuing activity may, in their judgment, exceed the net advantages to be gained from doing so. Fortunately they have available to them the alternative of a widespread network of external programs of top-notch quality, the subject of Chapter 11.

11

External Executive Programs

Defining Executive Programs

Executive programs (both internal and external) can be defined by organizational level; by potential for advancement; by content; by focus; and by teaching methods.

Organizational level: middle managers (manager of managers and up), upper-middle managers, top functional managers, senior executives and corporate officers. Titles include marketing manager or director, superintendent, operations manager, division manager, vice president, group executive, senior vice president, president, chief executive officer (rarely), among others.

Potential for advancement: corporate executive development programs are usually reserved for those managers and executives who appear on succession planning tables (normally the top 10 percent of the managerial group). Typical categories include high-management-potential middle managers (HMPs), replacements for incumbent executives, and incumbent executives.

Content: as outlined in Chapter 9, the content of executive programs is primarily directed toward business strategy; the external environment of business; and the internal environment of business, including the major functional disciplines and organizational behavior.

Focus: executive programs tend to be educational in nature, rather than a training experience whereby job skills are the primary concern. Updating knowledge, enhancing perspective, and reevaluating attitude are the objectives.

Teaching methods: in executive programs, the classroom process rests heavily on the experience of the executive participants (which is the primary learning vehicle), with the faculty taking the role of facilitators and providers of the latest research findings in their fields of expertise. Methods include lecture and discussions, case studies, small-group work, role playing, simulations, and special projects, among others.

Categorizing Executive Programs

While the categorization of executive programs is somewhat arbitrary, the following classifications should be helpful for understanding and differentiating among the various types of external executive programs.

General Management Programs

The objective of general management programs is to prepare executives who have functional backgrounds for future general management responsibilities, with particular emphasis on the integrative role of the general manager.

The typical curriculum would include a number of topics from the following list: business policy and strategy; U.S./international business environment; management systems; management-science/quantitative methods; management techniques and styles; organizational behavior and human resources; economics; finance and accounting; marketing; operations; and industrial relations.

General management programs for the middle manager tend to focus on functional knowledge and management techniques, while the programs for senior managers normally contain material on business policy, strategy, and the business environment.

Functional Programs

Typical functional program can run from one to two weeks long. Their categories are described as follows.

FINANCE

Objectives: to enhance knowledge and understanding of corporate financial and accounting systems and concepts, and their key role in successful business operations, for both financial and nonfinancial executives.

Content: capital structure; capital budgeting; capital markets; financial statements; dividend policy; mergers and acquisitions; transfer pricing; financial and cost accounting; international finance; exchange rates; inflation accounting.

Focus: the more senior programs deal with strategic financial issues, such as capital markets and budgeting, mergers and acquisitions, and international finance. The shorter, lower-level offerings (usually for nonfinancial managers) provide a nontechnical overview of basic accounting and financial concepts, including financial statement analysis and accounting systems.

MARKETING

Objectives: to provide executives with current knowledge and state-of-the-art understanding of the marketing function from a variety of perspectives.

Content: marketing strategy; pricing and growth strategies; market analysis; analytic tools; organization of the marketing function; sales planning; the distribution function; market research; product life cycles; sales force management.

Focus: middle management programs deal more with the sales and distribution function, while those programs designed for more senior executives emphasize the strategic aspects of marketing, including corporate and environmental perspectives.

OPERATIONS

Objectives: to increase understanding and provide state-of-the-art knowledge of the operations and manufacturing function of industrial organizations, including computer applications to manufacturing processes and automation.

Content: manufacturing and corporate strategy; production planning and materials management; inventory management; systems; CAD/CAM/CAE applications; process control; robotics; human resources management; QWL; production management; procurement; functional integration; capacity planning.

Focus: the more senior programs would cover the strategic questions in the operations sector, such as corporate strategy relationships, productivity, quality, and application of new technologies. Middle management programs deal more with the operational aspects of manufacturing, such as production control, materials control, human resources development (HRD) issues, and inventory management.

TECHNICAL

Objectives: to enhance understanding and up-date the knowledge of managers in engineering, research, and technical areas of the role and impact of technology on business operations.

Content: designed for either the management of R&D and engineering organizations, or the technological and scientific content and implications of decision making. The following topics would be covered depending on the type of program in question: R&D and corporate strategy; R&D interfaces with marketing and operations; technology trends; R&D organizational structure, staffing; innovation; leadership issues; current developments in the hard sciences; social and legal issues; project management.

Focus: more senior executives would tend to participate in the managerially oriented programs, while middle-level technical managers would likely enroll in offerings with heavy technical content.

Specialized Programs

Executive programs that do not fall neatly into the basic functional categories previously listed are typically from one to two weeks in length. Programs considered specialized can include the following topics.

HUMAN RESOURCES DEVELOPMENT

Objectives: to emphasize to human resources professionals the linkage of corporate strategy and human resources management, and to evaluate human resources policies and practices as they relate to organizational success in an increasingly competitive world.

Content: organizational change; human resources strategy; human resources policies and practices (selection and staffing, compensation, development and training, succession planning, appraisal systems).

Focus: attendees are usually experienced human resources professionals at upper-middle management levels.

LEADERSHIP AND ORGANIZATIONAL BEHAVIOR

Objectives: to enhance a positive behavioral change of managers in organizational settings through improved skills, increased knowledge and understanding of the leadership process, and more effective interpersonal relationships with subordinates, peers and superiors.

Focus and content: lower level offerings typically focus on content areas such as leadership styles; organizational development; team building; self-awareness; group dynamics; and communication skills. The more senior programs treat such topics as personality functioning; stages of psychological growth; psychological contracts; motivation theory; stress management; and managing change.

STRATEGY

Objectives: to increase the understanding of the strategic planning process and its role in national and international business for both line and staff executives.

Content: concepts of strategic planning; environmental scanning; corporate social responsibility; industry structure; competitive analysis; organizational issues; strategic planning and control systems; and corporate strategy interrelationships.

Focus: senior executives and top staff planners would be most attracted to strategy formulation programs, while middle-level managers would tend toward sessions dealing with the issues of strategy implementation.

PUBLIC AFFAIRS AND GOVERNMENT

Objectives: to update line and staff executives from the private sector on the operations of the U.S. federal government, public policy issues facing the United States, and private sector interface problems.

Content: social, political and economic environment; project and policy development; program review and evaluation; government functions and agencies; special interest groups; regulatory bodies; role of administrative, executive and judicial branches; fiscal and monetary policy; regulations; public/private sector challenges.

Focus: senior executives are more likely to enroll in programs with a public policy thrust, while middle managers tend to be more interested in the operations and functions of the federal government.

HUMANITIES

Objectives: to increase executive insight into the process of managing in an increasingly complex world of interinstitutional dependencies through the study of man, society, and values.

Content: justice, property and power; man and state; ethics; fairness and freedom; science and society; ideals and values; culture; alienation and motivation.

Focus: there is no necessary relationship between managerial level and topics covered in any particular program. Most humanities offerings utilize classic and contemporary literature as the basis for content delivery.

NARROW-FOCUS PROGRAMS

In addition to the major executive program categories already listed, there is another class of programs designed to meet the needs of executives and managers with special interests in specific sectors and dimensions of business activity. These programs tend to be short (a few days to one week) and are specially designed to meet specific interests.

Examples include programs for executives working in and with interests in the following areas: banking and finance; educational management; health care; mergers and acquisitions; real estate and construction; service and transportation; small business and entrepreneurship; utilities management; board of directors programs; and women's programs.

Degree Programs

Finally, we should recognize that executive education, whether provided by corporations on an in-house basis or by universities and other institutions on an out-company basis, is an essential component of the total development process for executive growth over the long term. All of the programs offered in the preceding categories play a key role in this process, and are, by design, nondegreed.

On the other hand, degree programs also play an important part in executive education. There are well over 100 executive MBA programs offered by U.S. business schools, as well as hundreds of standard MBA programs that can be taken on a full time basis or at night. The normal time frame is two years, with an MBA or some other master's degree awarded after successful completion.

Like the standard MBA program, the executive MBA program usually caters to high-potential middle managers, but the executive participant attends classes one day a week, plus one- to two-week residential sessions each year, while retaining full job responsibilities. The content of such programs is similar to that of the general management programs already listed. These programs require a significant sacrifice of time, effort, and investment on the part of the individual and his or her company and family.

It should be noted that there are two full-time degree programs that are nine months to one year long and residential, leading to a master's degree in management: the Stanford Sloan and MIT Sloan Programs. They are attended by very high potential middle managers, and require a one year leave of absence and a family move to the program location.

A footnote on degree programs: experience has shown that a formal degree has little or no relevance to long-term career success in a given corporation. Yet, an advanced degree can make a difference in the case of a young manager with intentions of moving from one company to another to gain career objectives. Apart from special interests in inter-company mobility needs, a formal credential in the form of an advanced degree is of limited value within a corporate career path. The name of the game is on-going executive education that can be integrated into the total development process in a systematic way to achieve the new levels of competence required by the executives of tomorrow.

To optimize the significant investment of time and money in continuing executive education, the various program offerings must be carefully analyzed to ensure an appropriate fit between needs and programs.

Program Differentiation

The major issue in selecting an external executive program is to find an offering that meets a specific and definable set of development needs for a given executive. Such needs are usually included in the individual development plan (IDP), which is a key document for all executives listed in most corporate succession planning systems.

This selection process requires an in-depth understanding of external executive programs—what they offer and how they differ from one another—so that a reasonable fit can be made between a given program and a set of defined needs. For example, in the general management program category, there are literally hundreds of offerings in the United States, with at least forty of these being the most prestigious products of the nation's leading business schools. Since they are all "good" or they wouldn't stay in business, how can their programs (and others) be differentiated?

Sources of Program Information

To gather information about a program, utilize any or all of the following sources:

- Feedback from previous participants can be obtained from evaluation forms completed by the executive participant and usually retained by corporate files.
- Feedback can also be obtained from corporate human resources development (HRD) or executive development staffs that make their evaluations based on regularly scheduled program visitations.
- Program brochures can be helpful in obtaining a general feel for what is being offered. However, these glossy products rarely provide any real basis for assessing the quality or uniqueness of that program.
- Executive program directories are a practical source of detailed information on most of the general management, functional, and specialized programs offered in both the United States and overseas. To be most useful, a directory should contain both evaluation as well as descriptive data on listed programs. Some publishers also offer phone-in consultation services on program selection.

Differentiation Criteria

Whatever the source of program information, a systematic approach to making comparisons among alternative programs should be developed. Such an approach should be based on the identification of major and minor criteria that can applied as a yardstick in the evaluation process.

Major criteria encompass the characteristics of a program's faculty, curriculum design and content, director, and participants. These important dimensions of an executive program provide the key to program quality.

Faculty. The strength of individual faculty members in their chosen fields of expertise is critical to program effectiveness. Currency and depth of knowledge, platform skills, teaching experience in executive programs, motivation and dedication to the learning process, and academic credentials are the principal elements of faculty competence. To achieve faculty strength, some general management programs look to other universities, to consulting firms, and to major corporations to provide the best resources. Other programs rely exclusively on in-house faculty who spend full time with the participants and work closely together to provide integration of the material being presented.

Curriculum design and content. The structure and subject matter of executive programs play a key role in determining the quality of the learning experience. This area is particularly crucial for general management programs that typically have eight or ten subject areas to cover in a program of a median length of four weeks. There are two basic approaches to curriculum design: blocked or streamed. The former consists of offering topics one at a time, whereby, for example, finance is covered for one or two days, followed by marketing, operations, organizational behavior, economics, and so on, with the final segment being devoted to integrative material on business strategy, policy, and environment. Streamed design consists of presenting the course material in parallel, whereby each day would include finance, marketing, operations, organizational behavior, and so on, culminating in an integrating theme in the final week. In addition, some programs utilize a combination of these two approaches. The objective is to provide a learning experience that is rational, integrative, and holistic, rather than the episodic, loosely connected array of topics that characterizes most MBA programs. The content of general management programs (previously listed) must have high credibility and currency, always reflecting the state-of-the-art issues and concerns facing executives all over the world.

Program directors. Those who plan, conduct, and evaluate executive programs should be dedicated professionals, with both business and academic backgrounds. They should be hands-on managers who pay close attention to participant needs as well as to the daily details of program execution. They are also responsible for maintaining close contact with the marketplace to assure that corporate executive development requirements are met.

Participants. As has been previously stated in Chapter 8, the primary learning resource for an individual in an executive program is the participant group itself. That fact underscores the importance of careful selection of executive attendees. The composition of the group is a joint responsibility of both the participants' parent organization and the program director. Serious consideration must be given to the managerial level, years of experience, motivation and maturity, and diversity of organizational backgrounds of applicants. General management program brochures frequently state the requirement of high growth potential for attendees. While most corporations use succession planning tables as the source of candidates (thus assuring, to some degree, this requirement will be met), such programs should also be open to senior people with major responsibilities and vast experience, and who remain at very high managerial levels even though they have limited future promotion opportunities. They can make a major contribution to the learning experience. The key is to assure appropriate peer relationships and a rich diversity of backgrounds to get maximum benefit from the educational experience.

While of lesser importance in contributing to program success, the following dimensions also play a significant role in the achievement of stated program objectives.

Teaching methods. While the case method, originally developed by The Harvard Business School, remains the basic teaching tool in many leading executive programs, we are seeing a movement toward a variety of methodologies to better accommodate a range of specific learning objectives. A lecture-discussion format may be the best way to present economics and the business environment; role playing is probably the most effective way to deal with leadership and organizational behavior issues; and computer-based business simulations can be very powerful tools in dealing with strategy and functional integration. Small group discussions are almost always used in preparation for classroom case study work, which sharpens analytical skills and problem-solving competence. In summary, a blend of teaching methodologies can most effectively meet each of the unique learning requirements in a well-planned executive program.

Facilities. Classroom, housing, and food and other services are peripheral factors, not being central to program selection. However, when the quality of the learning environment is not of a high standard, participants can be distracted by the lack of air-conditioned rooms, by uneven food quality, and by inattentive staff. To make sure that the program environment is a supportive and appropriate setting for executive education, many of the leading business schools have invested millions of dollars in new classroom and residential structures. Such schools include Northwestern, Michigan, Duke, Virginia, Wharton, and Babson. Others offer their programs in luxurious resorts or elegant conference centers, such as Columbia's Arden House, and Boston University's offering at the Ocean Edge Conference Center. Increasing competition in the field of executive education indicates the need for further attention to and investment in facilities.

Program length. One of the sticking points in selecting an executive program is deciding how long it takes to deliver a learning experience that is acceptable to both buyers and sellers. From the customer's perspective, (typically a corporation), the shorter the better because busy, successful executives prefer to be away for as little time as possible. From the program management's and, particularly, the faculty's point of view, the longer the better, because they believe it takes time for the educational process to become effective.

The program length issue primarily concerns general management programs, which can run from one week up to twelve weeks in length. In most cases, four- to five-week offerings are acceptable to most corporate customers. While they do participate in the more prestigious, longer programs (such as Harvard, Stanford, MIT, Virginia), the majority of executive attendees favor the shorter programs that can deliver a first-rate educational experience in that time frame. In recognition of corporate needs to do more in less time, there is a definite trend toward reducing the length general management programs (see Appendix C).

Program scheduling. Another approach to maximizing the effectiveness of the learning experience and, at the same time, minimizing time away from the job is to offer the program on a split-session basis, typically with a series of one-week modules separated by a one-month period back on the job to allow for application of what was learned. The advantage to the split session is the integration of learning with actual experience. The disadvantage lies in committing busy executives to return several times each year to complete the program. Also, such split-session programs tend to be regional in nature, where a multisession commitment can be made more easily and travel expenses can be minimized. Examples of some of the better known split-session programs include Northeastern, North Carolina, and Tennessee.

Another version of split-session offerings is to spread the learning experience over two or three years, typically in the summer. In some instances, a business-oriented project is assigned to be completed prior in the final session. Smith, Wabash (humanities program), and Ohio State are good examples of this design.

Program cost. The charges for tuition, materials, housing, and food are rarely a factor in selecting an executive program unless such costs are are greatly out of line. In 1991 fees for general management programs ranged from $2,100 to close to $4,000 per week, with a median of approximately $2,800 weekly. The median weekly charges for functional and specialized programs, which tend to be shorter by averaging about two weeks in length, run somewhat higher. These costs do not include travel or executive salaries. It is to be expected that program fees will rise in the foreseeable future, at about the annual inflation rate or a bit higher.

Special features. To differentiate their product from competing programs, most universities and other institutions offer a variety of special features that add to the educational experience.

Spouse program: The vast majority of general management programs provide an opportunity for spouses to participate, usually during the final week. Options include sitting in on regularly scheduled classes, participating in a specially designed series of spouse sessions, or a combination of these options. Humanities programs, particularly those for more senior levels of management (as at Aspen and Dartmouth), welcome spouses to join their husbands or wives for the entire program. More often than not, companies will fund spouse attendance providing real participation occurs and not merely attendance at end-of-program social activities.

Physical fitness program: In recognition of growing importance of executive health in the United States, increasing numbers of general management programs offer optional sessions in physical fitness. They are usually cardiovascular in focus, and include a physical evaluation by professional staff prior to participation. In addition to the benefit of offsetting the long hours of sitting in a classroom, such programs can lead to a meaningful follow-up exercise regiment after return to the work environment. The majority of executives enthusiastically volunteer for these fitness sessions and many continue to job and exercise after graduation from the program.

Business simulations: The majority of the leading general management programs offer some form of business game, with about an even split between computer-based and non-computer-based simulations. These exercises, which replicate the real world of business decision making, can be used to achieve a range of program purposes, including strategy formulation and implementation, negotiation skills, tradeoff management, management systems, organizational behavior, and management science. The primary contribution of a business simulation is in providing integration of course content through a pressure-packed, highly competitive series of realistic experiences that involve all of the managerial skills needed to be successful back on the job. While simulations tend to be scheduled during the final week of the program to pull it all together, an alternative design is to thread the simulation through the last several weeks of class work, permitting immediate application of course content to the decision process. Such simulations provide a risk-free opportunity to try out new behaviors, to apply innovative management tools, and to manage a business in a competitive environment. As in the case of jet pilots who are required to spend many hours in a flight simulator each year, "you don't have to run into a mountain to become more proficient"!

Other special features: The directors of external executive programs also include a number of other enhancements to strengthen the educational experience. Examples include microcomputer labs, electives, behavioral styles workshops, outward bound experiences, and alumni programs.

The Program Selection Process

Once sufficient information has been gathered about the various programs available, the selection process is undertaken to assure a proper fit between a program's offerings and the needs of the prospective attendees. For best results, this process should involve a number of people, from both the corporation seeking to utilize the program and the establishment offering it.

Role of the Corporation

INDIVIDUAL EXECUTIVE

The manager or executive who will be attending an external executive program should have a key role in the program selection process. His or her interests, needs, and career aspirations must be factored in to the decision as to which program is most appropriate at a particular point in time. In many cases, the initiative for choosing a program is taken by the individual executive who has very limited information on program offerings. Program brochures are of marginal value in differentiating the unique features that are of critical importance in selecting the right program to meet the special needs and interests of that individual. Frequently, the potential attendee utilizes the "old-boy" network of past participants from that company who have participated in a particular program and who found the experience very rewarding. "My boss went to Cornell and recommended it highly!" While Cornell offers a fine program, this approach to program selection ignores a number of alternative offerings that should be evaluated to meet the special needs of the individual. As previously mentioned, executive program directories can be extremely useful in providing detailed information to help in the selection process.

IMMEDIATE MANAGER AND BUSINESS UNIT EXECUTIVE

While a prospective attendee's "chain of command" superiors may possess information to aid in the selection process, their primary role is to approve the selected program and provide the necessary preattendance briefing with the assistance of corporate HRD and executive development staffs.

EXECUTIVE DEVELOPMENT STAFF

Having responsibility for coordinating companywide executive education, the executive development staff is usually located within the corporate personnel or human resources functions. If the organization offers internal executive programs, this staff, in conjunction with the corporate office, designs, executes, and evaluates such programs, particularly if they are held in a company residential facility dedicated to this purpose. In instances where programs designed specifically for company-oriented educational objectives are subcontracted, this staff would handle the negotiations with the business school selected, including requirements for content, faculty, and off-site facility location. It would also evaluate the effectiveness of results.

The executive development staff in many large corporations also has the responsibility for administering corporate use of external executive programs. This includes maintaining a central data base on programs, for those that have been attended by company executives as well as those that might be utilized in the future. Central to this information-keeping function is the maintenance of an evaluation file, with inputs from recent attendees and from program visitations by the staff. This program information is particularly valuable in assisting line management in the program selection process, by suggesting alternative offerings to meet defined developmental needs. Finally, through close coordination with the corporate office, the executive development staff establishes a corporate executive development strategy (see Chapter 6), linking the scheduled use of both internal and external executive education with on-the-job development requirements identified at various career points for those executives on succession planning tables.

It should be noted that under conditions of increasing decentralization, external program selection and administration is more frequently found at the business-unit level with coordination arrangements with the corporate staff.

SUCCESSION PLANNING STAFF

A succession planning staff may report to corporate human resources or directly to the corporate office. It is responsible for the establishment and maintenance of the succession planning tables (described in Chapter 4) that include the key executive resources of the corporation—high-management-potential middle managers (HMP's), incumbents, and replacements for incumbents or, as previously stated, about 10 percent of the managerial group. These tables are regularly reviewed by the corporate office to assure timeliness and accuracy.

The succession planning staff coordinates its activity closely with the executive development staff because it provides the "bodies" to attend the executive education programs identified by the latter staff. Personal data contained in the individual development plan (IDP) provides the primary source of executive development needs, and serves as the basis for a proper fit of needs to program offerings. Thus, through the integration of a business unit's succession planning with that at the corporate level, readiness, potential, and probably next assignment can be identified for various candidates (including all incumbents, replacements, and HMP's at the business-unit and divisional levels as well as those at the corporate level) so that the appropriate educational experience at the right time can be specified. These procedures are being increasingly computerized with uploading and downloading capability so that systemwide integration can be achieved.

After inputs from the succession planning and executive development staffs are provided, final approval of program selection is usually made by a top division or business unit executive. Only in the case of the two Sloan programs (MIT and Stanford), which require a substantial investment by the company and nine months to a year away from the job, as well as a physical move for the family, does the sign-off take place by the corporate office or the chief executive officer.

Role of the External Program's Director

The director of an external program is responsible for providing corporate and other clients with a clear understanding of what the program is designed to deliver, its unique features, its faculty strengths, its objectives and methods. Apart from the design and execution of that educational experience, the director must see to it that the class mix is appropriate, that the managerial level is consistent, and that the curriculum has adequate flexibility to meet the varying needs of executive attendees from a broad range of organizations both from private and public sectors. To assure that the market is being served, most directors travel extensively to personally interview corporate and other decision makers as well as, in some cases, the prospective attendee.

U.S. Establishments Offering Executive Programs

Colleges and Universities

There are at least 100 U.S. universities whose business schools offer executive programs, and, of these, 40 provide the leading general management and functional programs attended by most U.S. and international corporate executives. Examples of the most prestigious schools include Harvard, MIT, Stanford, Columbia, Virginia, Michigan, and Northwestern, among others. Some universities offer executive programs under the continuing education department. Examples include Stanford (humanities and functional) and Chicago (organizational behavior).

Apart from university business schools, there are a number of U.S. colleges that offer executive programs. Examples include Dartmouth College (general management and humanities), Williams College (humanities), and Babson College (functional programs).

Institutions

In addition to universities and colleges, there are a number of privately endowed institutions offering executive programs in various categories. Examples in the areas of public affairs and government include the Brookings Institution and The Washington Campus. In the organizational behavior category we can list The Levinson Institute and The National Training Laboratories (NTL). The best known in the humanities field are the Aspen Institute and The Dartmouth Institute. In the functional area, there is the Institute for Advanced Technology and The International Marketing Institute, among many others.

Government Agencies and Departments

The federal government, and its many agencies and departments, is heavily involved in executive education, and offers a broad range of programs in general management, organizational behavior, and others of a focused nature to address the special development needs of the agency in question. A few of these include the Federal Executive Institute (FEI), The Postal Institute, The Department of Agriculture, and The Office of Personnel Management (OPM). In addition, the Defense Department is noted for its development programs for military officers through the major war colleges and the Industrial College of the Armed Forces (ICAF).

Professional Associations

Many trade and professional associations provide executive programs for their members, with a special emphasis on the development needs of both members and other executives on an invitational basis. Perhaps the best known is the American Management Association (AMA), which offers a broad range of seminars and programs in a variety of categories. Other examples include The American Association of Certified Public Accountants (AACPA), American Banking Association (ABA), The Life Office Management Association (LOMA), and The Conference Board (CB).

Consulting Organizations and Others

Consulting firms, in addition to, or as a part, of their normal range of consulting activities, offer executive programs both on an in-house as well as a public basis. Such leading firms as Harbridge House, Arthur Anderson, and the Arthur D. Little Management Education Institute are good examples.

A number of other organizations that are engaged in research also are well known for their executive education programs. Two examples in the organizational behavior field include The Menninger Foundation and The Center for Creative Leadership (CCL). (See also Appendices B and C.)

European Establishments Offering Executive Programs

The leading international general management programs are found principally in continental Europe and the United Kingdom. These programs are somewhat unique in that they tend to be truly international in character (international faculty, international attendees, and an international curriculum), particularly those in continental Europe. Another distinguishing feature is that they are largely independent entities, with a limited affiliation with local universities or business schools (usually in terms of their MBA programs for certification purposes).

In the U.K., the leading institutions offering executive programs (as well as a number of consulting, in-house subcontract programs, and research activities) include The Ashridge Management College (a private, not-for-profit trust), Cranfield School of Management, Henley Management College, The London Business School, Manchester Business School, and the Templeton College Oxford Center for Management Studies. Most of these institutions are semi-independent with some industry, government, and university affiliations.

In continental Europe, the leading institutions include INSEAD (originally affiliated with the Harvard Business School but now independent), and two other organizations that have recently merged: International Management Development Institute (IMEDE) and The International Management Institute (IMI). The new combined entity has been named The International Management Development Institute (IMD). Both INSEAD and IMD are foundations, engaged in research and in supplying in-company training in addition to public programs.

Finally, there are two organizations deeply involved in management development activities in Europe. The EFMD (European Foundation for Management Development), which is a network of institutions and individuals devoted to executive education, and The International Management Education Consultancy (IMEC), which provides informational services and consulting services on management and executive development programs. (See also Appendix C.)

12

Program Evaluation and Application of Learning

Measurement and Evaluation

The Return on Investment (ROI) Issue

The eternal question raised by users of executive programs, whether internal or external, is "What will I get in return for investing time and money?" The issue of value received by this considerable investment in executive education is one that deserves a great deal of attention.

To rephrase the question we can ask: "What measurable positive changes in on-the-job performance can be directly attributable to the program experience?" Over the past several decades, dozens of studies have dealt with this issue, many of which are largely testimonial in nature. Others (and all too few) deal with the evaluation question in a much more substantive way, utilizing experimental and control groups, to measure statistically significant positive changes back on the job resulting from the program experience. These studies are treated in detail in Appendix B.

The problem in measuring and evaluating executive programs lies in large part in their general nature, particularly those in the general management category. According to participant testimony, the value received is said to be "broadening," "renewal," "increased self-confidence," "new knowledge and perspective," and "skill enhancement," among other comments. Almost every executive attendee states that "it was a great experience and I learned a lot." But the real question concerns whether that experience has a positive impact on performance.

Measuring the impact of a learning experience is no simple matter. Some enlightened corporate executives and executive development staff

124

directors sidestep the issue of evaluation, stating that outcomes are a matter of faith, and investment in executive education is akin to that in research and development, both having a long-term payoff. Probably most using organizations subscribe to this point of view, although the need for further research is clearly called for.

Apart from formal studies, many organizations that make extensive use of external executive programs do seriously attempt to determine the payoff for both the individual attendee and for the company. Usually the effort entails in-depth interviews with the executive, and his or her subordinates, peers, and superiors. In one instance, a West Coast aerospace firm came up with one observation worthy of note: "Executives who attend these programs ask better questions." Not proof of positive change, but an interesting concept.

It may be worthwhile to comment on the evaluation question in terms of the program categories that have been previously described. The best-known and most-widely attended are those in general management, which deal directly with managing a business: management process, functions, systems, organization, and environment. Clearly, these programs are designed for direct relevance to executive work and business needs.

On the other hand, another class of executive programs deals with the more intangible aspects of leadership. These are programs in the humanities and the liberal arts that deal with the broader questions of man and society, which have been thought about by those in leadership positions for over 2,000 years. Participants in such programs are exposed to philosophy, ethics, literature, and issues of culture and governance, all with a focus on the qualitative aspects of man and society in an increasingly complex world. Even more so than in general management programs, the attendee has an opportunity to broaden horizons, to assess personal values and life, to reevaluate career goals, and, in short, to become a better person. And that better person may just become a better executive!

Consequently, measurement and evaluating executive programs is a tricky and largely unresolved concern. Participant assessment is a related and controversial issue in an educational environment.

The Participant Assessment Issue

Executive development programs, be they internal or external, are designed to deliver an educational experience in a low- or zero-risk en-

vironment. Such an environment permits open discussion, new behavior
experimentation, disagreements, and challenge to new or traditional
ways of thinking without the fear of assessment or evaluation by superiors or the corporate succession planning staff. The idea that education
and assessment can coexist in the classroom is highly controversial,
with the vast majority of program directors and academic professionals
being violently opposed to the contamination of the learning environment by any kind of assessment activity. This aversion to assessment is
particularly true for external executive programs, the directors of which
receive inquiries from time to time about how participant "did" to elicit
some form of feedback on program performance. As a matter of policy,
such requests are not honored and any further inquiries are firmly dis-
couraged.

To the best of our knowledge, no corporate internal program is currently involved with assessment, but there are two well-known instances
of assessment in major corporate residential executive programs that
were carried on over a period of ten to fifteen years and finally discontinued. It might be helpful to briefly describe these two examples to
analyze the various aspects of this practice and the reasons for its
discontinuance. We have chosen not to identify these two large companies, which continue as leaders in their fields and which have had and
still have national and international reputations for excellence in the
field of executive education.

CASE STUDY 1
We first describe the situation involving a large, multinational corporation with two major internal programs at both the middle- and senior-
management levels. Both programs were four weeks long. The assessment feature was introduced in the early 1960s, with three principal
components: staff evaluation, a peer impression project, and a series of
intelligence tests (discontinued after two to three years).

Staff evaluation data was collected during the four weeks, based on
classroom observations and on informal participant discussions at
meals, breaks, and social occasions. Staff input was summarized and
collated for the program director on the Friday afternoon after the participants completed the program. The basis for assessment was the
perceived potential for general management resulting from a series of
behavioral variables exhibited during the program experience. This data
was reviewed by the staff psychologist for consistency and validity, and

later forwarded to the corporate succession planning group for inclusion in the personnel file for each participant.

The peer impression project was originally developed by the military to identify leadership potential. This assessment tool is based on the assumption that your peers can make a reasonably accurate evaluation of your performance in an environment of close contact over a specific period of time. In this case, the four-week classroom experience was deemed sufficient to generate peer impressions under conditions of an intense program design, including a range of learning methodologies: small group discussion, role playing, and a business simulation, among others. The project was introduced at the end of the third week, with each participant evaluating all other class members through an instrument containing a number of behavioral factors related to managerial success. The data was collected, summarized, and evaluated by the staff psychologist, who scheduled one-hour feedback sessions with each participant during the final three days of the program. In addition, the psychologist was required to forward the peer results, along with the staff data, to the corporate succession planning staff for inclusion in the personal jackets of each participant.

The rationale for including assessment in the educational program was based on the concept that attending an internal executive program is an assignment like many others in the corporation, such as participation in task forces and special projects, all of which should be, and are, subject to an evaluation of performance in that role.

When the program first started, the participants were not told of the assessment component on the grounds that such knowledge would have a negative impact on the learning experience; they would be "playing games" to look good to the assessors. However, within the first two or three days in the classroom, under the pressure and intensity of the program, the participants quickly put aside any concerns over assessment and devoted full energies to coping with the challenges of the learning experience. Years later, it was decided to announce to the class at the opening session that the program included participant assessment, and the data would be forwarded to corporate headquarters for inclusion in their personnel files. It was also pointed out that such data was not primary but supportive, to reinforce (it was hoped) the on-the-job appraisals that play a major role in determining future careers. Also, the class was told that they would have the benefit of peer feedback, which should be of great value to each participant's self-development.

When the assessment part of the program was first set up, the staff included professional educators, psychologists, senior corporate personnel executives, and outside consultants, who would all assure credibility and objectivity in the evaluation process. Such staff members had dual responsibilities in the program: observation of participants both in and out of the classroom, and frequent teaching sessions in their area of expertise. Later, a new policy brought in, on a rotational assignment, young middle managers from the major functions to take on this tricky staff role as a supplement to the professional group. While acceptance of these short-term functional managers was not a problem in the middle-management program, there was a potential credibility issue in the senior program. However, the careful selection of these people as to maturity, judgment, and experience offset their relative youth and lower managerial level.

A carefully blended mix of subject matter and methodology was designed to provide a range of managerial behavior sufficiently varied for observation purposes, such as business simulations for decision making under the pressure of time constraints, small group discussions for leadership qualities, and provocative speakers for conceptual ability.

Participant roommates were changed each week according to a preset plan, to allow each attendee to have maximum exposure to all other class members. Meal seating arrangements were planned on a rotational basis, with a staff member assigned to each table. Also, a staff member was required to sit in on each classroom session to both observe speaker performance and participant behavior.

Assessment activities were eventually stopped because the participants were becoming deeply concerned about the potential negative impact on their careers, to the detriment of the learning process.

CASE STUDY 2
Another large multinational corporation established a residential executive program at a resort facility in the Far West. In the mid 1960s the director visited the executive programs of the company previously described to study the structure and content of these offerings, with a particular interest in the assessment feature. A senior management program was subsequently patterned after what had been observed. Class size was limited to twenty participants to permit effective assessment operations.

A staff psychologist was hired to put in place both a peer-impression and a staff-evaluation program, which he personally administered. Unlike the previous example, all teaching was conducted by external faculty, with no other professional or rotational staff to handle the staff inputs. The psychologist alone was responsible for both categories of data, which was reviewed with the director before being forwarded to corporate headquarters for inclusion in the files of each participant.

It should be pointed that the staff psychologist was required to sit in on all discussions by top management regarding the possible promotion and career potential of each executive who attended the program. In this case, the power of one individual to affect executive careers in that company was enormous—truly a "king maker"! This controversial program was continued many years after the program described in case study 1 was terminated. However, the entire executive program, along with assessment, was subsequently canceled, largely a result of deteriorating business conditions.

USEFULNESS OF ASSESSMENT IN EXECUTIVE PROGRAMS
Despite the many valid arguments against including assessment in an executive educational program, a case could be made for using the classroom experience to make predictions concerning further managerial success. Three dimensions are worth consideration.

1. *Leadership.* Significant patterns of informal leadership can be noted during the programs, particularly in small group discussions, role playing exercises, and business simulations.
2. *Cognitive or intellectual ability.* How a participant deals with provocative ideas, new concepts, and intellectual challenges laid out by leading consultants and academic authorities can be observed in the classroom.
3. *Power relationships.* In many major internal programs, top executives in the company play an important faculty role in leading executive discussions and making top-level functional presentations. Despite no formal chain of command or authoritative atmosphere being present in these classrooms, seeing how a participant handles himself or herself when dealing with a top executive can be revealing. The kinds of questions asked or the quality of responses are indicative of sensitivity in dealing with power figures.

Thus, these three dimensions of effective management are part and parcel of program experience, and can be observed in the classroom setting.

Assessment Versus Feedback

The preceding two case studies show how assessment was used in two internal general-management executive programs. In one of these, individual feedback was provided within the assessment activity, although the primary purpose was to provide input to corporate succession planning tables.

Another approach to assessment and feedback is through the use of assessment centers (see Chapter 5, under Prevention of Derailing). In recent years, the trend has been to use assessment centers for developmental feedback to program participants, particularly in other large corporations. Typical methodologies include in-baskets, role-playing exercises, and simulations. Thus, key elements of managerial behavior are identified and shared with the participants. Some well-known external programs, such as the National Training Laboratories (NTL), have over many years offered various forms of sensitivity training that provides direct feedback within a small group structure. Thus, developmental feedback seems to be the major thrust in executive programs that may or not have an assessment feature.

While program evaluation and participant assessment are important issues, the fundamental issue is the individual payoff from an educational experience or, more important, from a series of such experiences over a total career. As pointed out in previous chapters, the "name of the game" is self-development and life-long learning, and the ultimate responsibility for growth lies with the individual executive. In his book, *The Human Side of Enterprise,* Douglas McGregor (1960) helps put this in perspective

> There is a real danger that the pressure for evaluation may lead us to try to measure the wrong things and, therefore, to miss the true value of experiences of this kind. The purpose of most of these generalized university courses is not, and should not be, direct practical application of the learning to the job. Their purpose is not to provide answers to problems, formulas, or tricks of the trade. It is to broaden the manager's understanding of his job, to challenge some of his preconceptions, to make him better able to learn from experience when he gets back home because he will have acquired a more

realistic understanding of the causes and effects with which he must deal. To the extent that this kind of education is successful, it will not reveal itself in immediate or obvious changes in his behavior back home. The learning which takes place will probably be reflected in fairly subtle ways of which he himself may often be unaware. Nevertheless, these changes in perception do affect behavior, sometimes profoundly. It is certainly reasonable that management should want to evaluate the achievements of university programs in management development, but it is important that we understand the purposes of these programs so that we evaluate the right things.

Reentry and Application of Learning

Executive education programs purport to deliver change, in knowledge, skills, attitudes, and perspective. A comment by a leading speaker in a major in-house executive program defined the application problem in the following way: "If each of you were to apply what you have learned in this classroom upon your return, your company would grind to a halt in 24 hours!" Is your organization ready for changed behavior? What will be the impact on subordinates, peers, and superiors?

To answer these questions and to further assure that the investment in an executive education program is worthwhile, it is clear that the responsibilities of the individual executive, the corporation, and the program director don't end with the completion of the course. All interested parties must be involved in smoothing the transition from the classroom to the business environment. Likewise, they must be active in directing the application of what has been learned for optimum return.

Corporate Responsibilities

ACCOMMODATE DEBRIEFING

The issue of program debriefing should be dealt with during the reentry period. The purpose of such a debriefing is severalfold: to provide an opportunity for management to understand what the objectives and content of the program were; to evaluate how effectively the program was in meeting these objectives; to determine what elements of the program experience are applicable to the organization's needs both in the short and long run; and to add information to the corporate data base of executive program usage. This debriefing should be done as soon after return to work as possible, preferably during the first week back on the job. It should take precedence over immediate job problems, including loaded in-baskets and pressing corporate issues.

Debriefing should be conducted by the participant's immediate manager, particularly if he or she had done the original preprogram briefing. Others in attendance could include the human resources or executive development director, as well as selected peers and subordinates. If such a meeting has not been arranged, then the initiative should lie with the executive participant. In view of the significant investment of both time and money by both the individual attendee and the organization, feedback and potential applications should be of obvious importance to the organization making such an investment.

PROVIDE AN APPROPRIATE CORPORATE CULTURE

There is an enormous variation in corporations in terms of values, "personality," openness, readiness to change, characteristics that reflect typical aspects of corporate "culture." Some companies might feel that there should be an immediate payoff to the investment in an executive program. This issue would likely be a central concern in the debriefing session, to see what should be applicable to the job situation. For example, a new approach to inventory control or the latest development in management information systems presented in the program could be carefully reviewed in terms or realities and needs of the organization. In one recent case, a linear programming computer model covered in the program was introduced by the participant into a corporate manufacturing facility with great success.

However, application of specific techniques learned is not usually the obvious result of having attended a program. The new ways of thinking and innovative approaches to managerial problems that are more likely results may be resisted or viewed as a threat. The corporate culture may be such that overt change is not readily acceptable. In one sense, large corporations can be viewed as inertia systems that respond very slowly to new ideas, much like a 500,000 ton supertanker trying to turn around quickly. As we look into the future, the restructuring of corporate America may give rise to the increasing need for creativity, openness, and flexibility in corporate culture, thus providing a greater receptivity to the real potential contribution of executive education programs.

VISIT PROGRAMS

Many companies sponsor visits to executive programs to get a firsthand impression of program quality. Such visits are usually the responsibility of executive development staff members, the participant's manager, or other company executives. The objective is to get overall participant feedback on the program experience and to obtain a personal impression of the operation. These visits usually last from one to two days, and generally involve sitting in on classes, participating in informal conversations with students, and meeting with faculty members and administrators. Being evidence of corporate interest, these visits are welcomed by the program management and provide a realistic assessment of the program.

Program Director Responsibilities

In addition to providing a meaningful and relevant curriculum, an excellent faculty, effective learning methodologies appropriate for experienced and mature executives, conducive facilities, and sensitive staff support, the director of executive programs should place a high priority on the transfer of learning to the participant's work environment. It is not enough to mount a fine program and forget about application back on the job. "Love 'em and leave 'em" is too often the case, particularly in general management programs in the United States. A number of actions can be taken by the program director to minimize the feeling of abandonment.

Provide a student log. Give each executive a series of guidelines in notebook form to record key ideas from each session and how they might be applied. Toward the end of the program, this diary, or log, can, on an optional basis, be reviewed with a faculty counselor for comment and suggestions.

Plan final-week small group discussions with peers and a faculty member consultant to review a major business issue experienced by each individual participant and to get feedback on the solution, with particular emphasis on how elements of the program might be applicable.

Plan for reentry. Arrange for the participant to meet with an appropriate staff member to discuss tactics and strategies for reentry and application of what has been learned in the program. This approach could include a session on how to develop results in an "Action Plan," with a suggested format dealing with career objectives, job challenges, peer and boss relationship issues, and specific plans for implementation of relevant program material.

Provide follow-up. The director should maintain periodic contact with program graduates to assess application successes or failures, to review problems and challenges in application, to provide counsel and support if appropriate, and to obtain feedback as to what changes and improvements should be made in the program to facilitate application.

Participant Responsibilities

In the final analysis, it is the individual executive who bears the primary responsibility for smooth reentry into the workplace and the application of what has been learned. The critical goals for the returning executive are to first identify what has been learned, and then, more important, to carefully plan how to introduce these ideas and concepts to the job situation in such a way as to assure better managerial performance with minimum disruption to the organization.

DEVELOP A REENTRY STRATEGY

It has been said that the strategy for reentry is to return as if never having been away; that is, show no overt changes in behavior or attitudes, no strange ideas or bizarre concepts, no new and threatening ways of dealing with management problems facing the business. But change can be introduced in such a way as to minimize disruption. In one case, an executive participant was "turned on" by a particular approach to problem solving and decision making, and was determined to apply these concepts back on the job. Unfortunately, his associates were baffled and confused by the terminology and ideas involved. Realizing that he was getting nowhere, he decided on an indirect approach and introduced the new technique in a subtle and nonthreatening way in later meetings where business problems were being discussed.

In the program setting, particularly in the final week of classes, the individual should start thinking about reentry if he or she hasn't done so earlier in the program. Questions that should be raised include

- What have I learned?
- What is my plan for applying this learning?
- What am I going to do differently?
- What is the potential impact on my peers, subordinates, and superiors? How will I deal with this?

Returning to the real world of the corporate work environment can be a traumatic experience. The individual will face piled-up mail, full in-baskets, unanswered phone calls, problems requiring immediate attention, checking in with the boss, and attending to colleagues and subordinates. And, despite such pressures, some time must be found for debriefing, for sharing what was learned and exploring ways of application.

DEFINE LEARNING OBJECTIVES

Starting with a preprogram briefing by management as to program content and objectives, the individual attendee should at that point try to define what learning objectives he or she has, what specific areas of needs have been identified, and how the program can meet these needs. In addition, what current business problems might be solved by insights gained from the program experience should be noted.

Some general management programs require that participants bring a personal or business project to be worked on during attendance. In many cases, these projects have been identified in advance with faculty consultation, worked on during the program, and followed up by faculty advisors. This approach has the advantage of providing a meaningful and practical focus to the learning experience, utilizing relevant subject matter and faculty counseling to achieve a real-world payoff for both the sending corporation and the individual executive. However, some general management program directors take a different point of view, stating that projects cause the participant to become too preoccupied with the assignment and too internally focused, which works against the objectives of such programs, that is, against broadening and externalization.

The issue is whether both purposes can be served: the need for practical application of learning to solve immediate business problems on one hand, and the stated objectives of general management programs on the other hand: to provide a broadening, stretching, renewal experience in an environment of new ideas, with the freedom to explore alternative approaches to managing and to gain insights on career goals through interaction with a rich group of participants from a broad range of organizations and an experienced and competent faculty.

13

Educational Alliances

Executive programs are increasingly a joint responsibility of both the universities and other institutions supplying executive education services on the one hand, and the corporations, government and other not-for-profit organizations that utilize these services on the other. Yet, as we intend to show, the line of demarcation between buyers and sellers of executive programs is becoming blurred.

More and more, new cooperative arrangements between business schools and corporations are being created to deal specifically with the development needs of executives and managers from a wide spectrum of users. There is no longer a clear-cut distinction between internal programs and external programs, as there are now a range of opportunities for executive education reflecting the growing interdependence of buyers and sellers in delivering more focused offerings. As previously stated, executive education is being viewed as a joint responsibility of university program directors and corporate human resources staffs, working together to plan, execute, and evaluate this activity to help assure that the product will be timely, relevant, and appropriate for executives who will be facing the complexities of the next century.

Public Programs

In the United States there are at least sixty major suppliers of public programs, including such well-known names as Harvard, MIT, Dartmouth, Aspen, Menninger, Brookings, Virginia, and Columbia, among many others. These programs provide open enrollment offerings by business schools and other institutions in general management, functional and specialized areas. In recent years, over 15,000 executives annually from the United States and abroad have enrolled in such programs, and the number of participants will be steadily growing in the future. One of the issues facing the directors of these programs, particularly in general management, is how to provide sufficient flexibility in curriculum design to meet the wide spectrum of needs represented in a typically diverse set of participants. Several schools are addressing the tailoring issue through optional sessions and case studies in the service as well as manufacturing sectors. Overall, public programs meet a special need for a general educational experience by a group of executives from a wide range of backgrounds, dealing with the common managerial challenges facing most organizations in the international business environment.

Consortia and Partnership Programs

A step removed from public programs, consortia and partnership programs have limited enrollment and are designed to meet the specific needs of two or more organizations. They can be held on-campus or at company facilities. One of the two best-known consortia programs is the Babson Consortium, which is offered in two modules of three weeks in June and two weeks in August, with nine participating companies. Current member companies include AT&T, Digital Equipment Corporation, Eli Lilly and Company, DowElanco, North American Philips Corporation, Norton Company, New England Telephone, Perkin-Elmer Corporation, and Textron Defense Systems. This program grew out of a need for a vehicle through which member companies could influence program design and evolution, yet provide exposure to other corporate cultures and management practices. The second program is a fairly recent consortium put together by Amoco in partnership with Indiana University for eight Fortune 100 companies. Again, the concept is to provide a focused program to meet the needs of a restricted number of companies.

New England Power and Light has developed a partnership for the public utility industry in a program conducted by the University of Michigan. An example of a two-company partnership was established in Europe between IBM-Europe and Shell UK. This program was designed to meet the needs of two large and somewhat different companies with an equal number of executive participants from each. It utilized faculty from both corporations as well as from leading European Universities. There are many other examples of such cooperative arrangements both in the United States and Europe.

Subcontract Programs

The growth in demand for company-specific executive education exceeds that of public programs and is likely to continue to do so into the future. As a result, subcontract programs conducted by universities have been designed to meet the needs of a single organization. The bulk of these programs are held on campus to take advantage of academic surroundings (although company facilities can be used), are typically one week in length, and provide a general management experience in most cases. Advantages to the business schools include intellectual stimulation for the faculty in focusing knowledge and experience on specific corporate purposes and incremental income to the school. From the corporate perspective, reduced educational costs can be significant and pressing corporate issues can be addressed. Also participants from the same organization can strengthen their company's culture.

An increasing number of business schools are providing subcontract programs, with recent surveys showing over forty U.S. business schools providing such services. On the other hand, a number of leading institutions (including Harvard, MIT, and Stanford) do not undertake to serve individual companies; they don't want to divert faculty resources from their public executive and degree programs.

In-House Programs

As do subcontract programs, in-house executive programs are designed to meet the needs of one organization. However, the difference between the two types is that in-house programs are planned and conducted by corporate executive development staffs and are usually held in company residential facilities. University faculty consultants provide sessions in their areas of expertise in conjunction with modules led by corporate executives and executive development staff members. Among the most widely recognized prestigious in-house programs are those of IBM, General Electric, GTE, Xerox, Motorola, Lockheed, and Hewlett-

Packard. With the possible exception of GE, these companies also utilize public programs to meet the overall development requirements of their executives.

In-house programs tend to be expensive, with considerable investment in "bricks and mortar," as in GE's Crotonville facility and IBM's Management Development Center. It is likely that such investment in corporate facilities will be minimized in the future, and that subcontract programs held on university campuses will play an increasing role in serving the market for company-specific executive education.

Thus, we see a spectrum of executive education programs, ranging from public offerings and consortia/partnership cooperative arrangements, to company-specific programs conducted on both a subcontract and solely in-house basis. (See also Appendix C.) These categories of programs are summarized in Table 13.1.

Conclusion

Each category of executive education program has its own particular advantages, moving from a general education in management to more tailored and focused programs designed to meet specific company needs.

As previously mentioned, an ideal corporate executive development strategy would require attendance in both internal and external programs on a regularly scheduled basis throughout a management career based on specific needs at a given point in time. In this way, life-long education can play a key role in assuring continuing executive competence in the years to come. To bring this about, both universities and corporations are developing closer links in marrying research, teaching, program design and execution. Thus, the linkage between universities and corporations is serving to meet the increasing demands of the business community to maintain vision, productivity, quality and international competitiveness in a rapidly changing world.

TABLE 13.1   U.S. Executive Programs: General Management, Functional, and Specialized

Program Category	Characteristics	University Role	Organization Role
Public	Open enrollment (all organizations) General focus and some tailoring	Program design and execution University campus	Selection of participants
Consortia and partnership	Limited enrollment (select organizations) Designed to meet the needs of two or more organizations	Program design and execution University campus facility	Input to design and execution Selection of participants Optional company facility
Subcontract	Company-specific (one organization) Designed to meet the needs of one organization	Program design and execution University campus facility	Define company needs Input to design and execution Selection of participants Optional company facility
Inhouse	Company-specific (one organization) Designed to meet the needs of one organization	Provide academic faculty Occasional assistance in program design	Define company needs Program design and execution Selection of participants Provide company facility and some teaching resources

EPILOGUE

Several observations or lessons can be deduced from the account of executive development and education we present in this book. For one, it is evident that two broad strands of development need and growth are interwoven throughout the years of an executive's career. Each of those strands is recognized in the many programs conducted by universities and corporations.

1. On the technical side of management, the increasing complexities of operating a business under accelerating rates of change, in a tight financial and competitive environment, and on a global scale, impose ever shrinking margins of safety in business planning, timing, practice, and performance. And so the executive must be continually reassessing and refurbishing his or her operational business knowledge and skills.
2. On the personal side, and probably even more significant than the technocratic aspects of management, lies the need for the executive to enhance his or her competencies in business leadership and vision, recognizing the multiple internal and external forces that impact with increasing force on any organization or institution as it moves toward a new century of global involvement.

In both these aspects of management new orders of excellence are called for. Ever-diminishing margins of competitive advantage are increasingly decisive in the drive to maintain a healthy and growing enterprise.

Other lessons include the need for awareness of the following facts of survival for both business and its executives.

- Leadership in the twenty-first century will become ever more critical to the success of the organization.
- Executive incumbents must exhibit some of the attributes of the so-called universal mind; their mental horizons in the exercise of their profession must be universal in scope.
- To develop a universal mentality, incumbents must grow through-

141

out their careers as whole persons. Technical expertise alone won't make the grade.

• The symbiosis between the executive and the organization will become increasingly important to the survival of each.

From the foregoing lessons we can draw two conclusions. One is that society as a whole must look to the continuing lifelong growth of its leaders with the same kind of fervor and concern that it addresses to the problems of educating its children. To any solution it must contribute three ingredients: an understanding of the need, a common intent, and the will to act. In any case, society will always get the kind of youth or leadership that it deserves.

The second conclusion involves the expanding role of the university. Executive education can no longer be only an adjunct, a side show, to the university curriculum or to its historical functions of precareer or early-career education. Lifelong and formal educational offerings must be integrated into the spectrum of the university's contribution to society. The university must operate in partnership with organizations in every walk of life, each of which shares responsibility for developing society's leaders. That is to say, the university must share in life, not stand detached from it.

Still other lessons lie in the future. But we who stand always on its threshold can discern some of its features, as outlined in the earlier chapters of this book. Trends point to a spreading throughout all of humanity of new social values; of pragmatic release from old dogmas; and of continuous learning and growth for individuals throughout the middle-class populations of the more developed nations of the world. Just possibly we'll see a world at peace with itself if enough people have the wit and will to make it happen. The great opportunity lies just ahead, but it's not there for the taking. We must earn it.

The activities we call executive development and education are in step with what is happening on that broader stage of social and economic history (see Appendix C). Such education is consonant with this general dictum: that the purpose of living is to learn and the purpose of education is to help us learn more from our living. For some portions of the population and in some walks of life that philosophy has always been true. It is now being made explicit for the leaders of organizations, an imperative for dynamic new societies as they evolve throughout the nation and the world.

And so we see that personal development can no longer be regarded as only for the young and well-to-do among either individuals or nations. It is becoming a necessity among all nations throughout this threatened world of ours. In America we are recognizing our manifest need for a much better educated and developed general work force. The leadership of that work force likewise must face up to the challenges of its common future. The foregoing accounts describe what is being done to help it meet those challenges. What is being done is a good beginning.

Appendices

Appendices

APPENDIX A

Predetermined Things

The following material is an attempt to identify those trends of change throughout society that are so basic and pervasive as to be instrumental in describing its future over at least the next few decades (forecasting). The effects of those trends are what we mean by *predetermined things*.

The process of distinguishing between predetermined things and uncertain things is an essential step in generating scenarios about the future. Yet for anyone thinking objectively about such matters, for you as well as for us, the task still remains to try not to confuse familiar events or conditions with the underlying trends they supposedly portend.

We raise that caution because, as you read the following recitation of changes that we think will continue to be a significant part of the future, their interpretation finally rests on value judgments that you can make at least as well as we can. We can only suggest the input data. Nevertheless, being aware of the risks faced in forecasting, while not accepting uncritically what you read here, can help you to see more clearly how the past may indeed be prologue to the future.

One kind of risk is that of unawareness. The origins of change exist but are hidden from us long before their consequences become evident. For instance, we now see that the application of worldwide instantaneous visual communication has changed the way nations fundamentally think about and deal with one another on both the diplomatic and popular levels. Twenty years ago we could speculate, but not likely predict as a predetermined thing, the extent of that outcome simply from the invention of the cathode ray tube.

Another risk lies in the dynamic instability in the course of change. For instance, does the breakup of the USSR as we have know it increase or decrease the chance of nuclear warfare? It could go either way; it all depends. Is it going to rain tomorrow? It all depends.

A third kind of risk results from the swiftness with which change can occur. In order to communicate we have to use contemporary terms and events that can rapidly become obsolete even though the underlying trends remain intact. For example, in earlier drafts of this material, a few years ago, we wrote of the drift toward regionalism in political and trade relationships, and used contemporary illustrations from events in the Middle East, China, and the Soviet Union to illustrate that thesis. Subsequent events in those regions have made those references sound absurd. But the underlying trend toward regionalism held then, does now, and arguably will continue into the plannable future. No doubt this material as you read it contains some equally dated references and language because of the rapid pace of change in today's world.

We can only ask the reader to be aware of such problems in forecasting the future, and to try to visualize how the underlying trends of change throughout the nation and the world will also change the demands on the executives of the future. In fact, the mere existence of those problems in forecasting supports the thesis of this book: the need for life-long learning and growth of the executive. For it is on the ability to understand the past, to be aware of the present, and to read the future that their own futures will greatly depend.

Social and Cultural Trends: Significant Changes in Population Growth, Aging, and Work Force Makeup

In 1975, world population stood at 4 billion people. By the year 2000, the projection is for 6.35 billion, an increase of over 50 percent. This projection amounts to 100 million births each year. By the year 2030, the Earth's population is estimated to be 10 billion, and is projected to reach 30 billion by the year 2100, which, according to experts, will be the Earth's limit! Of this growth, 90 percent will occur in the poorest countries, resulting in a rather constant GNP/capita of $200/year in Asia.

By the year 2000, the ratio of retirees to active workers in West Germany will be 36 percent; in Japan, 30 percent; and in the United States, 24 percent (the older population is increasing twice as fast as the general population). In the year 2010, one out of five Americans will be 65 or older. Every day in the United States, 5,000 people turn 65, and three out of four people who reach 65 will live on the average until they reach 81.

The implications for the U.S. social security system are significant. Currently, the system is holding up because the baby-boom generation is working and putting money into it. However, in the next twenty years when this group starts retiring, the system may not fare so well. Social security taxes have had to be increased eightfold since 1970 to keep the system afloat. Currently, 3.4 working people support each retiree: but when the baby boomers retire and are followed by a smaller group of young people, there will be fewer than two workers for each retiree.

In terms of labor force makeup, 25 million workers will be needed to provide the general population with goods and services by the year 2000, according to a Hudson Institute study (Silk, 1980). Most of these will be nonwhite, female, or immigrant workers. Native white males, who now constitute 47 percent of the labor force, will account for only 15 percent of the entrants to the labor force in the year 2000. Put another way, by the year 2000, 25 percent of U.S. workers will be black or Hispanic, with most new hires being women, blacks, and Hispanic or Asian men. More women in the work force will mean 75 percent of families will be dual career in the year 2000 versus 55 percent now.

Environmental Trends: Major Concerns in the Physical Environment

Energy

During the 1990s, world oil production will approach geological estimates of maximum production capacity, even with increases in petroleum prices, according to the U.S. government study, "Entering the 21st Century" (Barney, 1980). The richer industrialized nations will be able to command enough oil supplies to meet rising demands through the end of the century. Even with a modest rise in prices, the less developed countries will face increasing difficulties meeting their energy needs. While the world's finite fuel resources—coal, oil, gas, oil shale, tar sands, and uranium—are theoretically sufficient for centuries, they are not evenly distributed, and pose difficult economic and environmental problems.

For the one-quarter of humankind that depends on wood for energy, the outlook is bleak. Needs for wood will exceed available supplies before the turn of the century. Significant losses of world forests will continue as the demand for forest products and fuel wood increases. Growing stocks of commercial-size timber are projected to decline 50 percent per capita. The world's forests are disappearing at the rate of 18 to 20 million hectares a year (an area half the size of California) with most of the loss occurring in the humid tropical forests of Africa, Asia, and South America. The projections indicate that by the year 2000, some 40 percent of the world's remaining forest cover will be gone.

Nuclear power may be the long-term solution to the ever-increasing demand for energy. However, we see a moratorium on building new plants in the United States (Shoreham, Seabrook, and others in the northeast) due to extensive cost overruns and concerns for safety, despite the extensive use of nuclear power in France and many other countries in Europe. A whole new look at this issue seems inevitable as we approach the twenty-first century and realize that the limits of fossil fuel energy supplies are being reached and that burning fossil fuels is ravaging the environment, whereas nuclear power does not contribute to ozone depletion, to acid rain or to the greenhouse effect.

Pollution

Contamination of the world's seas and atmosphere seems to remain a serious problem. Atmospheric concentrations of carbon dioxide and ozone-depleting chemicals are expected to increase at rates the could alter the world's climate significantly by 2050. The greenhouse effect may already be upon us, as the extraordinary drought conditions in the U.S. mid-west with higher than normal temperatures have caused nationwide concern since 1988.

In 1988, medical waste along the U.S. shoreline caused the closing of many beaches in the northeast. Boston harbor is said to be one of the most polluted bodies of water in the country. Offshore dumping of waste is under serious review, and increasing efforts are being made to modernize treatment facilities throughout the country. Acid rain from increased combustion of fossil fuels (especially coal) is causing damage to lakes, soils, and crops. Radioactive and other hazardous materials present health and safety problems in an increasing numbers of countries.

One of the biggest environmental battles ahead will be over water. Sewer and water capabilities are far from adequate in high-growth areas. Water supply is a problem in a number of regions in the United States: northern and southern California, Arizona, Colorado, and the

Great Lakes states. Major aquifers of the Midwest lie beneath fields that farmers have treated with pesticides for thirty years. All of these factors indicate that an obsession with water will be with us into the next century.

Political Trends: Significant Shifts in the World Political Environment and U.S. Public Policy Issues

Decline of Communism and Rise of Democracy

The revolution that swept eastern Europe in 1989 might be considered one of the most dramatic events of the twentieth century. Communist dictatorships in Poland, East Germany, Czechoslovakia, and Romania have been overthrown and supplanted by democratic regimes dedicated to capitalism and free markets. While the struggle with communism is largely over, the struggle for the long-range success of capitalism is just about to begin. The transition to free markets and free societies where they have not existed before is proving painful and may be impossible in some cases. The reintegration of the two Germanies, starting with open borders and now a common currency, has exciting possibilities both politically and economically but will pose many problems, particularly in East Germany, such as rising prices and unemployment.

The most explosive development in the decline of communism has been the dissolution of the Soviet Union as a nation in January, 1992. Under Boris Yeltsin, Russia has become the centerpiece of a new federation of former Soviet republics that is still in its formative stages. Yeltsin's attempts to introduce radical economic reform in Russia and the other republics has met with serious resistance, and the prognosis is that there will be long term instability in that region. Facing and effectively dealing with these challenges will be the the number one priority for the world's leaders.

In terms of disarmament, President George Bush's proposal to unilaterally cut the U.S. arsenal of both tactical and strategic nuclear weapons stunned the world as that initiative would be, for the first time, a significant step to remove the threat of nuclear war. Yelsin's reaction has been more than responsive in matching or exceeding Bush's reductions. While recognizing that the former USSR's nuclear arsenal is physically located in Russia and several other former Soviet republics, a series of agreements as to how this downsizing will be accomplished will provide a formidable challenge to the United States and those republics.

Regional Instabilities and Relationships

With the proliferation of nation-states (160 represented in the Korean Olympics, and 159 in the United Nations) nationalism and regional conflicts seem inevitable. The Middle East situation seems incapable of solution, with a beleaguered Israel on one hand and a large number of relatively unstable Arab states on the other. U.S. strategic interests in the area are clear, with the oil resource viability being the basis of policy. This issue may be resolved in the long term once the United States develops a coherent approach toward energy usage and conservation. While making some progress towards development and democratization of government, Latin America continues to be predisposed to revolution and political instability. As the debt load eases and more responsible leadership emerges, we may see real growth and social progress in that area.

The United Nations, which for years has been viewed as relatively ineffective in attaining its original goal as a forum for resolution of world conflict, is changing. Today the United Nations is receiving full support from both the United States and other major world powers. Recent accomplishments by the United Nations include:

- Negotiations between Afghanistan and the former U.S.S.R. established a timetable for the withdrawal of Soviet troops from that country.
- In July 1988, Iran announced acceptance of a U.N. resolution for a cease-fire between Iran and Iraq.
- U.N.-mediated talks with South Africa, Angola, and Cuba led to a cease-fire and U.N.-supervised elections for Namibian independence.
- Morocca and Polisario guerrilla forces in the Western Sahara accepted a U.N. plan to end 13 years of war.
- Solidarity was achieved in the United Nations during the 1991 Gulf War.

Another encouraging development at the United Nations reflects a decline of anti-Western rhetoric as Third World nations begin to face up to their own political and economic shortcomings. While recognizing that these successes may not be entirely attributable to U.N. efforts, the increasing maturity and responsibility of this organization bodes well for the future under the leadership of the new Security-General, Boutros-Boutros Ghali.

Trade and Protectionism

The 1986 declaration in Punta del Este, Uruguay, by the world's leading trading nations launched a new round of world trade talks designed to combat protectionism. In the fall of 1988, a textile bill was passed by the U.S. Congress limiting the total imports of textiles and apparel that has caused a great deal of consternation in the European Economic Community (EEC). The EEC felt that this new trade law would lead to increasing protectionism in the United States, thus causing it to renege on its obligations under the General Agreement on Tariffs and Trade (GATT), the set of regulations governing world trade.

According to Walter Wriston, retired CEO of Citicorp, history confirms that not only do controls fail, but they distort the global marketplace. The World Bank has documented that if someone gains a dollar from protection, someone else in the same country loses a lot more. For every $20,000-a-year job in the Swedish shipyards, Swedish taxpayers pay an estimated $50,000 annual subsidy. Protection slows economic recovery. To the degree that imports are cut, costs are increased to the consumer and is retaliation invited. The only way the world can sell more abroad is to buy more abroad.

There are those who feel that America is floundering in the global marketplace, incurring significant losses in market position, profits, equity, and jobs. The problem seems to lie not so much with U.S. products but with our trade policy, which was developed in the 1940s when Great Britain and the United States dominated the world economy. It is also based on the assumption that world commerce would be conducted under the terms of GATT. Today, only 7 percent of global commerce is covered by GATT, while over 75 percent is conducted by economic systems other than Anglo-American. It seems that the multilateral world requires more options and flexibility, both inside and outside of GATT, and a reconciliation of the free trade and fair trade advocates.

U.S. companies got back into the world marketplace in the last quarter of the 1980s, inspired by a lower dollar and strengthened by corporate restructuring. While imports remain strong (and probably will continue so into the foreseeable future), much of the export growth has been in the traditional areas of aerospace, scientific instruments, paper, and chemicals and pharmaceuticals, with many of the companies in the small and medium-size categories achieving a greater share.

The rest of the world produces four times as much as the U.S. GNP. Of the Earth's population, 95 percent lies outside of the United States and is growing at 70-percent faster rate. According to a recent McKinsey survey, 70 percent of the aggressive U.S. midsize companies expect their international sales to increase 15 to 20 percent a year during the early 1990s. Historically, U.S. companies have been myopic about exports (5.4% of GNP in 1988 versus 26% in West Germany, 25% in Canada, and 10.5% in Japan), due partly to the broad and very rich national market. Fortunately, this viewpoint is changing with the conclusion on the part of increasing numbers of enlightened U.S. companies that the future lies in treating the world as a market without borders.

A key development having a significant impact on world trade will be the full integration of the twelve-member European Community scheduled for January 1, 1993. A common currency, no tariff barriers, and an economic and political union with a passport-free mobility among member countries will characterize this new integrated Europe. A vital issue is how the Community will deal the rest of the world in terms of Japan and the United States having access to this vast new market.

During the 1990s, a steadily improving trade balance is virtually certain, due in part to a major rebound in the volume of merchandise exported and a peaking in volume of imports. The falling dollar and increases in production efficiency also contribute to an improved trade balance. U.S. manufacturing productivity has been on the rise and has exceeded West Germany and matched that of Japan in 1988. This trend should continue and should also relieve pressures for strongly protectionist trade bills, which the U.S. Congress will continue to consider.

The U.S. Deficit, Interest Rates, and the Dollar

U.S. President Ronald Reagan was probably right in 1981 when he said, "A society, like an individual family, cannot live beyond its means indefinitely." Unfortunately, we in America continue to see federal spending at about 23 percent of GNP with federal taxes at only 19 percent of GNP. This surge in indebtedness has increased the federal debt from $1 trillion in 1980 to $2.5 trillion in 1988, and $3 trillion in 1990.

Therefore, one of the major policy issues facing the United States during the remaining years of this century, and probably into the next, is deficit reduction. Depending on which expert you talk to, the deficit projection for fiscal 1991–1992 will reach $348 billion. However, this figure is distorted by costs of the Gulf War and the savings-and-loan bailout. Over the next few years, the effects of these events will be minimized with deficit reductions resulting. Myriad proposals to reduce the deficit have been made, both to reduce spending and to increase revenue. They pose difficult choices to achieve the objective without increasing inflation, causing recession, impacting productivity, and reducing the American standard of living. Some likely candidates for expense reduction include paring down the defense budget (particularly strategic defense initiative research), and farm subsidies, and even taking a hard look at the sacrosanct entitlement programs such as Social Security, Medicare, and veterans benefits. On the revenue side, new taxes will likely, probably in the consumption area: "sin" taxes on cigarettes and liquor, gasoline and oil import levies, and possibly a national VAT (value added tax), among others.

One of the greatest impacts of a serious move to reduce the federal deficit would be an increased level of confidence worldwide in the maturing role and heightened responsibility of the United States in international finance.

The budget deficit and the trade deficit are aspects of the same phenomenon: too much consumption and not enough savings. Savings and investment should be encouraged to build a solid industrial base for future economic competitiveness. What are the chances they will be?

One of the positive developments in recent years has been the decline of the dollar, which has already had a salutary impact on the trade balance. The cheap dollar will continue to dissuade U.S. customers from expensive imports, while strengthening exports. Meanwhile, interest rates have remained reasonably low, providing a stimulus for investment. A declining deficit will make room for private credit demands to expand without pressing interest rates up, which should stimulate business spending for capital goods and inventories.

In spite of the seriousness of the deficit, it seems likely that slow but steady progress will be made to achieve long-term deficit reductions. Meanwhile, it should be pointed out that the real issue may be the ability to service the debt and not to retire it. Continuous U.S. growth appears likely to take care of that. The 1988 deficit of approximately $150 billion was 3.2 percent of GNP. The Congressional Budget Office projects a net deficit of 4 percent of GNP in 1991, dropping to 2 percent of GNP over the next five years. Thus, the debt of America may never be retired, but should slowly but surely be reduced to a manageable level by the year 2000.

According to James Giffin, vice president of Aetna Life, a broader interpretation of the U.S. trade and budget deficits points to the creation of a fiscal boost to the global economy while placing the United States as a net debtor nation for the first time since the early 1950s. By shifting part of the debt burden to foreigners, the dollar decline has repositioned our industry, permitting us to draw on the economic energy we have pumped into our trading partners. In short, we have helped make their economies stronger and more of an equal partner in the economic arena. Far from pointing to a decline in the United States, our strength will continue to grow consistent with the increasing viability of our trading partners in the world market. That globalization will diminish the differences among national economic systems. Americans in the future may be no richer than any other national group but should be better off than they are today.

Business Trends: Major Changes in Structure and Operations

Restructuring and Decentralization

Beginning in the late 1970s, American companies responded to an onslaught of international competition by undertaking a massive restructuring. U.S. companies have been rebuilding themselves from the ground up, erecting a sleek new architecture to replace the unwieldy structures of the past. The general aim of this restructuring effort has been to sharply cut back on costs and to make dramatic and durable improvement in long-term profitability and growth. Through the 1980s, well over half of the names on the Fortune 500 list of the 1,000 largest U.S. corporations have undergone some form of significant reorganization and downsizing. Gulf & Western had spun off 65 subsidiaries; IBM had closed three domestic plants; USX had shut down three steel mills; Jack Welch at General Electric had sold old unprofitable businesses and bought new ones, had closed 73 plants and facilities, and had reduced GE headcount by more that 100,000.

In addition to downsizing, U.S. business has been moving in the direction of decentralization since the 1950s. In light of the need to become more competitive and flexible, decentralization may become even more critical in the future. This modern form of organizational structure is characterized by decentralized operations and smaller centralized service and control staffs. Such organizations have traditionally had a large number of middle-management positions that have served as a source of top management candidates. Under the pressure of restructuring, many of these middle management jobs are being eliminated, which may point to a long-term problem in executive succession.

The ongoing reduction of middle-management positions has resulted in a widespread ailment: middle-management malaise, or "the leaner and meaner blues." The cause has been dismissal of more than a million Americans from management and professional positions during the last half of the 1980s. AT&T recently cut 11,000 management positions; American Airlines reduced its Chicago headquarters staff by 25 percent; General Motors reduced the number of managers and salaried workers by 25 percent as of 1990 and is planning to close dozens of plants and eliminate more than 20,000 additional jobs by 1993; Ford cut its salaried payroll by 20 percent in 1990; ITT slashed its workforce by 100,000 (44%) including a reduction of headquarters staff from 850 to 350. Put another way, between 1983 and 1987, some 600,000 to 1.2 million middle and upper-middle level executives with annual salaries of $40,000 and more have lost their jobs, and an additional 200,000 to 300,000 are expected to receive pink slips by 1990. This downsizing trend is expected to continue through 1992.

An increasingly popular route to greater efficiency is to cut corporate staff, that is, the collection of economists, planners, marketers, human-resources specialists, and analysts who sit at corporate headquarters to help line executives do their work. This move may lead to more freedom for business unit managers to manage their shops, with needed services obtained from their own staffs or from outside consultants and experts. Thus flexibility and efficiency may be obtained without the seemingly vast overhead attributable to swollen staff bureaucracies that have been characteristic of U.S. business over the past decades. The trimming and reshaping of American corporations is making unprecedented demands on senior executives and the dwindling number of managers in the middle, such that more must be done with fewer people and a necessarily more highly skilled management team.

The painful restructuring is not over. Manufacturing jobs will continue to atrophy, names over plant doors may not be American in the future, but survivors in basic U.S. industries should emerge as low-cost, quality-driven winners. In other words, a revolution in thinking about how to restructure corporations is under way. Many see it as being healthy and constructive, and believe that it should be permanent revolution.

Mergers and Acquisitions

In the early 1980s, the number of large U.S. corporate acquisitions grew at a rate roughly double that of the 1970s, with a dramatic average takeover rate of 2,100 per year. The dollar value of mergers and acquisitions between 1985 and 1987 exceeded $520 billion, ten times the value of mergers between 1975 and 1977. Seen from another perspective, businesses representing 5.5 to 7 percent of the total market value of all U.S. corporations will have disappeared through acquisitions in recent years. During the last years of the 1980s, the merger wave continued, with corporations, buyout specialists, and other takeover artists putting some $50 billion worth of deals on the table during a two-month period in early 1988. It is felt that such activity is here to stay, having become an ingrained way of doing business. Mergers and acquisitions provide a number of advantages: a method of dislodging inept management, a prime means of staying competitive, and an easy way to make money.

It should be pointed out that, unlike previous paper-driven mergers in the 1960s, the fundamental forces driving today's takeovers are economic, primarily because of increased worldwide competition: consolidation and expansion. Many companies are competing with corporate raiders to acquire relevant businesses and thus help assure more prosperity in an increasingly competitive global economy. In addition, the trend is away from diversification, which historically had been based on the premise that a good manager can manage anything. Today's CEOs are more concerned with "focus," with being able to manage a business in a related field where operational skills and knowledge can be applied to an acquisition.

In order to enhance their competitiveness, a large number of U.S. companies are, and will be, forming strategic alliances through partnerships, joint ventures, and other agreements. The growth rate of such ventures has been picking up since the beginning of the decade, from 6 percent a year to over 22 percent. A good example is IBM which has forty active alliances, including several major partnerships in Japan. It is also one of fourteen companies supporting Semateck, a new research consortium in Texas.

On a worldwide basis, multinational corporations are becoming more globally cooperative through a mechanism of global strategic partnerships (GSPs), a new strategic option that covers every sector of the world economy, from sunrise to sunset industries, from manufacturing to services. AT&T has signed accords with Olivetti and Philips, European leaders in information technology. Ford, GM, and Chrysler have consummated production deals with Mazda, Toyota, Suzuki, Isuzu, and Mitsubishi. Madison Avenue's Young and Rubicam 1971 linkup with Tokyo's Dentsu, the world's largest advertising agency, has led to plans to merge with European affiliates of Eurocom, the French market leaders. Other alliances are between large and small companies, many of which are suppliers who could benefit from global markets and greater strength of resources. Criteria for GSP alliance are

1. Two or more companies develop a common, long-term strategy aimed at world leadership.
2. Relationship is shared and reciprocal.
3. The partners effort is global.
4. The relationship is organized along horizontal lines.
5. The participating companies retain their national and ideological identities.

While corporate collaboration may be difficult to achieve and presents many problems, such arrangements are likely to be the wave of the future for globalization of business.

In the United States from a regulatory point of view, the Reagan administration had adopted a laissez-faire attitude toward takeover activity, with little or no invocation of the Clayton Act's Section 7 that gives courts the power to stop anticompetitive mergers in advance. There has been some congressional concern to regulate raiders by requiring prior disclosure and other reforms. It is believed that under future administrations any such changes will be marginal, affecting the price of such deals but not the logic; such deals may slow but will not stop. Takeovers are now a structural part of the corporate landscape, and they are likely to continue as long as there are companies that can be made more efficient.

Foreign Ownership of U.S. Business

In 1987, foreign buyers spent $40.6 billion for U.S. real estate and corporations, and nearly $17 billion of that was in Smokestack America. In early 1988, such buyers had already spent over $10 billion in the manufacturing sector alone, and there is no doubt that this rate of investment in U.S. industry and business will continue to grow. Even with the great availability of worldwide financial resources, it seems that cost is not so much the issue as is the opportunity to get into American businesses in case U.S. policy turns sharply protectionist. The Japanese case is worthy of note. From 1980 through 1987, direct Japanese investment in U.S. firms totaled $33.4 billion, or a staggering 611 percent increase from its position at the start of the decade! By 1990, Japan will outstrip both Britain and the Netherlands as the biggest direct foreign investor in the United States.

Foreigners currently own more than 10 percent of U.S. manufacturing including concrete and consumer electronics, and 46 percent of the commercial property in Los Angeles. Foreign banks make one fourth of all commercial loans, and foreign firms employ 3 million Americans. In 1988, Japan's Normua Securities purchased a $100 million, 20 percent stake in the new Wall St. mergers and acquisitions firm of Wasserstien, Perella, thus giving it entrée into the growing takeover market in the United States.

Is foreign ownership of U.S. business a problem for this country? Probably not in the long run, as investment knows no national boundaries. Such foreign investment creates jobs, increases the U.S. balance of trade due to a growing level of exports by U.S.-based foreign firms, and constitutes yet another step in the globalization of business.

Globalization of Business

One of the predetermined things guaranteed to occur is the increasing globalization of business, which will operate in an ever more interconnected and shrinking world, requiring necessarily quicker responses to overseas developments and even more decentralization with responsibility closer to the operating level. According to Sony Chairman, Akio Morita, the economic system is more and more like a single interacting organism that must correct such internal imbalances as trade surpluses in some countries and huge deficits in others; and political leaders must avoid protectionism as a short run palliative to protect local industry.

One of the key issues of globalization is that of product standardization versus customized or tailored products to meet various customer needs all over the world. Uniform products are cheaper to produce, but are likely result in a loss of business in the world marketplace. A good case in point is the Boeing 737 aircraft. It was successfully modified to meet the needs of Third World countries and, as their airlines grew, the 737 became the best selling commercial jet in history. IBM makes dozens of different computer keyboards, and Europe alone needs twenty. Both of these examples emphasize the fact that adaptability must be built in to the product. Worldwide marketing effectiveness depends on serving the requirements of a highly differentiated marketplace.

Global competitiveness problems are serious. To surmount them, businesses must be cost-effective and focus on customers and markets, and must constantly pursue and integrate state-of-the-art technology into products and services.

One way to maintain competitiveness is to move manufacturing offshore. Historically, U.S. companies went abroad to secure foreign markets or to obtain raw materials. Now they go abroad to buy or make products and components to ship back home. Offshore imports rose from $953 million in 1966 to $36.5 billion in 1986, primarily in automobiles, electronics, and textiles. For example, in 1986, United Technologies moved two diesel engine parts operations to two plants in Europe; in 1987, GM phased out A-body car production in the U.S. and moved it to Mexico.

U.S. companies are continuing to augment their overseas facilities. Capital expenditures by foreign affiliates of U.S. manufacturers rose from $12 billion in 1978 to $17 billion in 1986, and continue to grow. Companies like RCA, TI, and Motorola employ more than half of the 73,000 people in Malaysia working in the electronics industry, and sent back to the United States $300 million in semiconductors in 1985; this volume continues to rise. Offshore manufacturing has a number of critics who maintain that labor cost savings will eventually disappear and societal pressures to keep work at home will continue to mount, and that such moves are more tactical than strategic. However, in the long run, overseas sourcing of products and services is likely to continue as being consistent with the on-going internationalization of business.

How can companies inculcate a global mentality? One of the surest ways is to develop a truly international team of managers adept at working abroad. Companies like IBM have a policy of staffing overseas operations by nationals of that particular country who know the local business environment and culture. This concept, together with an extensive program of regularly scheduled rotational assignments around the world, helps assure a continuity of executives competent to manage in an increasingly complex and demanding global business environment. A recent survey of executive mobility overseas showed a pattern of frequent job and industry change, averaging three years per job and five years per company, with most executives indicating they expect to change companies again, particularly those working in firms outside the United States. Thus, company loyalty may be on the wane, giving rise to a professional group of peripatetic internationalists who will be responsible for the fortunes of global business operations.

The Rise of Entrepreneurship

Entrepreneurial enterprise, which began to take hold in the 1950s, seemed to come into its own in the 1960s. The Reagan administration of the 1980s gave further impetus to entrepreneurship in the United States by providing political leadership committed to capitalism that was individualistic, risk-taking, job-creating, and growth-oriented (a kind of reaction to the welfare state concept), thus relaxing individual bureaucratic restrictions and organizational constraints. As a result, innovation and change have become buzzwords. The increasing availability of capital by venture capitalists who willingly fund new and innovative ventures supports the growth of entrepreneurial activity. Such conditions favoring individuality and creativity have lead to the rapid development of new and exciting products, particularly in the high technology areas. Large corporations have recognized the value of unleashing professional and managerial resources within the normal constraints of control, and have created small, independent business units (IBUs) to develop new products and markets with minimum corporate constraints. This concept is now known as "intrapreneurship" in the corporate community.

However, one current theory states that there is an inherent incompatibility between entrepreneurs and managers. It points out that managers manage towards existing marketing and financial objectives with a given product line, whereas entrepreneurs have confidence that the things they supply will create their own demand. But even small entrepreneurial firms need the application of managerial concepts to new products, new problems, and new opportunities. Given the intensification of global competition, both small and big business will require entrepreneurial virtues and management competencies.

Given the American export surge in recent years, an increasing share of this growth is coming from small and mid-size companies new to the export business. This share is being aggressively exploited by small entrepreneurial companies. In California, Guild Wineries sells inexpensive table wines in cardboard containers to Denmark, shipping 12,500 cases in 1988, with 13,000 in the last three months of 1988 alone. California produces 50 percent of the country's fruits and vegetables, including 90 percent of all exports in those categories, with 60 percent of the export crops going to the Pacific rim. Amway, manufacturer of cosmetics, soaps, and vitamins, sells more product in Japan than in the United States. Entrepreneurship thus will play an undoubtedly significant part in the global marketplace, having benefitted from the revitalization of managerial competence.

Growth of the Service Economy

The service economy employs 76 percent of all workers in the United States. During the 1970s and 1980s job growth in this country resulted from the expansion of the service sector, with a surge of 26 million service jobs from 1973 to 1987—a 47 percent increase—compared to Britain's service employment rise of only 22 percent and that of West Germany at 15 percent during the same period.

The traditional service industry includes financial services, education, transportation, utilities, insurance, and communications, among others. Today, however, manufacturing firms are moving into relevant service markets, which may be a wise move in an economy where manufacturing has already declined to about 20 percent of the U.S. GNP, and because services will probably account for six out of seven new jobs over by the year 2000. U.S. manufacturers have also focused on quality enhancement with an increasing emphasis on superior customer service in order to remain competitive. The objective is to solve current and future customer problems by selling a service rather than just a product. The financing of customers' purchases is another form of adding profits through service. For example, GMAC finances dealers' inventories and provides installment loans to customers. GE, on the other hand, with $12 billion in assets, operates a general finance company involving only a small portion of GE products in financing commercial and home loans, dealer inventories, auto leases, and more. Many other manufacturers are creating credit subsidiaries, thus increasing profits and developing deeper and long-lasting relationships with customers.

Business services account for only one-fifth of U.S. exports, but increasing numbers of banks, insurance firms, accountants, lawyers, executive search firms, ad agencies, and architects are going overseas to service customers who are expanding internationally. Will this activity hurt the balance of trade? Not necessarily, since in the long run the repatriation of revenues that flow back from services delivered abroad will enhance export earnings.

Will the United States ever become a totally service economy? Apart from defense, government, and other areas of strategic concern, it might be possible. The developed nations have moved from agrarian to industrial economies, and ever since World War I, there has been a trend toward services. The pace of that trend has accelerated enormously and should continue into the twenty-first century. However, we are now seeing a resurgence in the manufacturing sector, such that a more equitable balance between service and manufacturing will probably occur during the 1990s.

Technological Trends: Increasing Role and Importance of Applied Science

Computers

We now stand at the dawn of an age of insight, a new era of understanding of how things work and how to make them work better. The computer is being transformed from a number cruncher into a machine of insight and discovery capable of looking into the future. Today's supercomputers can be used to forecast the weather a few days ahead currently, to predict the structure of new materials, or to simulate cosmic phenomena. By the beginning of the next century, hand-held devices will respond to handwritten and spoken inputs. Supercomputers a thousand times more powerful than we have at the present will be able to calculate electronic interactions in molecules to create new materials. Doctors will be able to check heart conditions by having a patient walk through a diagnostic machine. Voice recognition systems will allow dictation of inventory figures. Expert systems will allow credit managers to evaluate the risk potential of new customers. Networks will be developed to connect currently incompatible brands of mainframe computers, minicomputers, and personal computers.

As we approach the next century, the growth of in personal computers (PCs) will be phenomenal. The PC is now approaching the speed and power of big mainframe computers, and will become easier to use, particularly for senior executives, only 10 percent of whom are currently using PCs. It is estimated that the number of PCs in offices will quadruple to 46 million by the next century. The computer will emerge as a full-fledged management aid to coordinate the many daily tasks of corporate administration. (One computer program already available is called "Groupware," It utilizes the principle of a management interaction system, a series of interactions that can be sorted into offers, counteroffers, commitments, and requests. The system features an electronic calendar and mail network and an information data base available to the entire organization.)

Telecommunications

One of the most exciting developments in the next century will be the emergence of highly sophisticated telecommunications systems, linking the world by great computerized networks able to process voice, video, and data with equal ease. Developments in telecommunications and computers are converging at an opportune time, providing the application of tremendous computational power to the growing importance of the telecommunications industry, which is switching from analog to digital (the basic language of computers) transmission of signals.

Through the use of fiber optic cables, the capacity of existing telephone lines will be enhanced to allow the simultaneous transmission of voice, video, and data. Fiber optics will permit the telephone to become an integrated information appliance, featuring a large flat screen thus permitting visual phone conferences (consider the impact on business air travel). It will be able to send and receive documents and messages, to act as a full-size computer, and to provide access to broad range of information data bases.

One of the more exotic potential applications in future telephone technology is being developed in Japan; it will provide instantaneous translation of a voice into at least two languages, English and Japanese, for a start. Apart from such awe-inspiring new technology, telephone books in Europe are being replaced by small computer terminals that can find addresses, buy train and theater tickets, or send flowers. This system is currently in use in nearly 4 million French households and offices as a government subsidized experiment.

Automation and Robotics

One of the most important contributions to the increasing productivity of the manufacturing sector in recent years has been, in addition to just-in-time production systems and new approaches to quality control, a new set of technologies collectively referred to as programmable automation. Using the computer as a base, these applications include computer-aided design (CAD), computer-aided engineering (CAE), flexible manufacturing systems (FMS), robotics, and computer-integrated manufacturing (CIM). These advances have improved and will continue to improve all aspects of the manufacturing process: costs, quality, flexibility, delivery, speed, and design. A recent study of twenty U.S. companies using FMS systems showed a labor reduction of 50 percent and a product cost reduction of up to 75 percent. While effective application of these technologies will continue to pose a management challenge, the contribution of such techniques will continue to grow in importance into the next century.

In 1984, the number of industrial robots used in Japan was 30,000, while in the United States, the number was only 7,000. Today, robot usage in the two countries has increased as the need for production efficiency and cost reduction in a global business environment is inevitable. The burgeoning service sector has provided a new market for robotics. For the past several years, service robots have been at work in nuclear plants and under the sea. Today, robots are being applied to surgery, care of the handicapped, cleaning services, and security. Future applications include building and maintaining offshore oil rigs, construction work, attending hospital patients, assembling space stations, fighting fires, preparing fast food, and inspecting high tension electric wires. In at least three applications—for nuclear plants and undersea and space uses—the projections for cost savings in addition to human safety concerns will be a big plus.

Superconductivity

Superconductivity is based on the phenomenon that the flow of electrons is greatly enhanced by cooling the conductors they pass through, to the point that resistance disappears at extremely low temperatures. Recent breakthroughs in the field feature superconductivity at much higher temperatures through use of new, exotic materials. Forecasts of application by the year 2000 to computing and signal switching, to magnetic levitation and 300-mph trains, and to cheap electrical transmission may be overly optimistic. However, continuing materials research will move us in the direction of increasing future applications of this technology.

Biotechnology

Some of the most exciting developments for the future lie in the fields of biotechnology and genetic engineering, with particular application to medical diagnosis and treatment. Some experts feel that by the year 2000 it should be possible to prevent such autoimmune diseases as rheumatoid arthritis, multiple sclerosis, and insulin-dependent diabetes, that result in the body mistakenly attacking its own tissue. Even more exciting is the possibility that whole organs can be regrown instead of being replaced. Genetic engineers have been able to extract healing substances from the human body and to duplicate them, thus providing the basis for new and extremely effective pharmaceuticals. One of the most significant potential achievements may turn out to be the deciphering of the human genome that determines the order, content, and location of genes in the human chromosomes controlling the body's growth and well-being, thus opening broad, new avenues of medical diagnosis and treatment. Deciphering the genome could allow doctors to prepare a genetic printout of a baby at birth to spot susceptibility to various diseases. A molecule might be designed that would connect to cancerous cells and make them revert to normal. Thus, biotechnology may allow for the best medical care in any hospital in the world in the next century.

APPENDIX B

Research on Executive Education

In Chapter 12, the issue of measurement and evaluation of executive programs was addressed in some detail. The central concern was stated as one of return on investment by organizations enrolling executives and managers in both internal and external executive programs, the main question being, "What measurable positive changes in on-the-job performance can be directly attributable to the program experience?"

Since the mid-1950s, a number of studies have been conducted dealing with executive programs in an effort to cast some light on the value of executive programs to the sending organizations, to the individual participant, and, in some cases, to the program faculty.

The research summarized in Tables B.1, B.2, and B.3 is organized into three categories:

1. Descriptive studies focus on the differences, corporate use, trends, market share, selection criteria, content, and benefits of programs from an users perspective.
2. Testimonial studies concentrate on the value of executive programs to the individual participant and to program faculty. The data consists of attendee testimony: reactions, attitudes, and benefits from the program experience.
3. Analytical studies deal with behavioral and attitudinal changes in program attendees as determined by the use of experimental and control groups, and include statistical analysis of the resulting data.

The tables highlight the major features of each study, while detailed information on each is included in the References section of this book.

TABLE B.1 Descriptive Studies

	Business Week (Byrne 1991)	*Bricker Bulletin* (Billy 1991)	Vicere and Freeman (1990)	Fresina (1988)
Number of years covered	1	1	2 (1988–1990)	2 (1987–1988)
Number of programs covered	100+	243	Numerous	Numerous
Number of participants covered	7,000 surveys sent to HRD staff and participants	15,840	171 corporations	300 corporations
Purpose	To rank executive programs in terms of teaching quality, faculty availability, and relevance of material taught	Ranking of the top 20 business schools in the U.S. and overseas according to number of participants and participant weeks in their programs	Designating benchmark executive education trends and comparison of trends to those of 1982	To describe the array of executive education offered both by companies and universities
Methodology	Four-page questionnaire	Questionnaire	26-item questionnaire	Questionnaires and interviews
Results	Gives a ranking of leading executive programs as well as what respondents thought of executive education	Indicates Michigan first in terms of number of participants; Ashridge first in terms of number of participant weeks	Shows an increased level of in-house executive education and use of university programs	Indicates a growing commitment to executive education by major corporations; company-specific programs will experience greatest growth; trends toward the centralization of executive education in corporations
Comments	Its broader perspective from HRD staff contrasts to that of participants; provides some surprising conclusions about the perceived quality of many leading flagship programs	One of many periodic surveys of market share for executive education programs	An updated survey of executive education trends in both Fortune 500 U.S. and international companies	A timely study of corporate executive education practices

	GMAC/Battelle (Saari 1987)	Ingolls (1986)	Schrader (1985)	Vicere (1983)	Harbridge House (Gibson and Jones 1979)
Number of years covered	1	1	1	3 (1981–1983)	1
Number of programs covered	Numerous	Numerous	Numerous	Numerous	38
Number of participants covered	1,000 corporations	400 corporations	106 corporations 312 participants 31 program directors	183 corporations	80 participants 38 program directors
Purpose	To survey corporate use of university, residential, executive MBA, and short courses as well as company-related specific programs	Investigating how executive education is integrated into corporate strategy	Determining program selection criteria and benefits from program experience	To determine executive education trend in Fortune 500 companies	To determine executive program differences and perceived benefits
Methodology	Questionnaire	Questionnaires and interviews	Questionnaire	Questionnaire	Questionnaires and interviews
Results	Indicates 95% use of at least one program; average of 3.9 executive attendees per company; $2 billion expenditure on executive education; trend towards increased use of programs	Provides a formulation of five basic approaches to executive education; contingency model balances corporate needs for internal and external programs and strategic adaption	Shows selection criteria by program content and objectives; university reputation; and program length; Benefits described principally as "broadening"	Shows slow but positive growth in executive education, with major growth in internal programs; gives advantages of both internal and external programs	Five program categories are identified as well as perceived benefits of each
Comments	A broad and intensive survey of corporate use of four categories of executive programs	A novel and interesting approach linking executive education to corporate strategy development	An interesting study from three perspectives of program selection and benefits	An informative three-year study of trends in executive education in the Fortune 500 companies in the mid-1980s	One of the few studies by a leading consulting firm; data somewhat inadequate in dealing with program differentiation

TABLE B.2 Testimonial Studies

	DuJardin (1985)	Hawaii (Jenner 1982)	ABT-Carbide (Berman 1981)	Federal Executive Institute (Newland 1976)
Number of years covered	1	4 (1978–1982)	1	7 (1968–1975)
Number of programs covered	1	1	5	32
Number of participants covered	70	60	299	246
Purpose	To determine the long-term impact of a 14-week general management program	To measure attitude shifts of participants in Hawaii Advanced Management Program	To evaluate and compare program effectiveness of five major general management programs	To measure program impacts on executive performance after attending FEI
Methodology	Sample of nine groupings of married men from the U.S., using questionnaires and in-depth interviews	56-item pretest and posttest questionnaire	Four-part questionnaire plus school visitations and interviews	36-item questionnaire in four categories
Results	Indicates that participants perceived a positive impact on management competencies, but some frustration in application of learning	Shows little attitude change in participants; faculty less influential than peers	Before: participant selection and promotion criteria identified During: beneficial program elements stated After: organizational climate data revealed	Indicates that program objectives were met; shows lasting effects on job and high and low benefit groups; gives program improvement recommendations
Comments	An unusual focus on careers and values, with stated changes in personal lives and life transition	A unique comparison of participant versus faculty attitudes	A comparison of program effectiveness of five major general management programs	A careful study of program impacts; no hard data on executive performance changes

(continued)

TABLE B.2 *Continued*

	DuJardin (1985)	Hawaii (Jenner 1982)	ABT-Carbide (Berman 1981)	Federal Executive Institute (Newland 1976)
Number of years covered	4 (1971–1974)	3 (1967–1969)	1	1
Number of programs covered	63	2	3	38
Number of participants covered	367	211	150 (est.)	6,000
Purpose	To evaluate and compare effectiveness of a large number of U.S. executive programs	To compare effectiveness of MBA and executive programs at Northeastern	Determining impact of executive education attitudes of participants in three MIT programs: Masters, Sloan, and senior executive	Assessing effectiveness of university executive programs
Methodology	Questionnaire containing 20 attitude and 24 content areas	Two-part questionnaire	100-item questionnaire and testing	16-page, 48-item questionnaire
Results	Data on 75% positive attitudinal reactions; data on 75% positive content reactions	Shows little difference in participant objectives; both programs successful; MBA program had greater career impact	Indicates a statistically significant shift of participant attitudes towards those of the faculty	Gives overall favorable data; specific benefits in knowledge, and skill, self-confidence
Comments	Summary data missing; an incomplete and inconclusive study	An interesting and unusual comparative study of degree and nondegree programs	Focus on similarity of participant attitudes and faculty attitudes towards business through	First and largest study of its kind focusing on perceived benefits from executive programs

TABLE B.3 Analytical Studies

	Sparks (1979)	Watson (1973)	Penn/Bell (Viteles 1956)
Number of years covered	1	1	3 (1953–1956)
Number of programs covered	1	1	1
Number of participants covered	40 (est.)	30	113
Purpose	Measuring cognitive learning and impacts on leadership in the Houston executive program	Measuring program impacts from attending the Illinois executive program on managerial performance	To evaluate the University of Pennsylvania's humanities program
Methodology	Experimental and control group design using tests, questionnaires, and interviews	Experimental and control group design using a 40-item questionnaire	Experimental and control group design, using questionnaires and tests
Results	Shows statistically significant increases in cognitive learning, but no or some negative impact on leadership effectiveness	Shows statistically significant performance improvement by experimental group	Shows statistically significant changes in attitudes of participants in the program
Comments	One of the few research efforts based on the experimental method; some criticism on validity of findings	Best study of program impacts on performace; Was Watson's Ph.D dissertation	First analytical study of a humanities program; no data on performance impact

APPENDIX C

Trends in Executive Education

Growth of Executive Education in the United States

In 1990, it is estimated that over 16,000 executives will have attended university and other institutional general-management, functional, and specialized executive education programs. Another 50,000 will have participated in internal corporate programs, and over 3,000 will have attended more than 100 executive MBA programs offered in the United States.

As we look ahead, the steady growth of executive education forecast by a number of recent studies (see Appendix B) can be extrapolated to show what this market will look like in the year 2000.

- University general-management and functional programs: 22,000 executive attendees at an estimated cost of $278 million, or $12,600 per executive.
- Specialized programs offered by universities and other institutions: 3,400 executive attendees at an estimated cost of $16 million, or $4,700 per executive.
- Internal corporate programs: 60,000 attendees at a cost of $348 million, or $5,800 per executive.
- Executive MBA programs: 4,000 attendees with the better programs costing $36,000 and up for the two years, and with an average class of 40 participants; the total investment will be over $150 million.
- Data on short programs and seminars are not available, but it is estimated that tens of thousands of managers and executives will participate in these two- to three-day programs at a cost of $300 to $500 per day.

Overall, we are looking at a total annual investment in executive education approaching $1 billion by the year 2000, excluding salaries, travel expenses, and investment in educational facilities. Executive education has been growing steadily since the 1940s, and will continue to expand into the foreseeable future. Of course, business conditions will have some impact on these projections, but the message is clear: in an increasingly complex world of international business activity, investment in executive resources is critical to competitiveness and corporate survival.

Globalization of Executive Education

Overseas Executive Programs by U.S. and European Institutions

There are, as of 1991, at least twenty executive programs conducted at overseas locations, most of which are in the Far East (see Table C.1). The most active U.S. institution is The University of California-Berkeley, which offers five programs in the Far East and one in the Caribbean. In Europe, INSEAD conducts five programs in Singapore and Macao dealing with Asian management problems. Other U.S. programs in Europe include Aspen seminars, Boston University programs and Virginia and Brookings offerings. Multicountry sessions include those by The University of Pennsylvania and Michigan's special program in global leadership.

International Participation in U.S. Programs

GENERAL MANAGEMENT PROGRAMS
As of 1991, there are seventeen U.S. general management programs that enroll at least 20 percent of their participants from overseas (Table C.2). They typically attract senior management from large organizations, run from four to eleven weeks, and are offered by the largest and most prestigious U.S. business schools.

FUNCTIONAL AND SPECIALIZED PROGRAMS
There are 27 major functional and specialized programs in the United States as of 1991 enrolling at least 20 percent of their classes from overseas (Table C.3). Functional offerings in finance, marketing, operations, and technical aspects include sixteen programs. Specialized pro-

grams in human resources development, organizational behavior, and strategy number eleven offerings. Mostly senior management attend these classes given by large, well-known business schools.

INTERNATIONAL PROGRAMS OFFERED AT U.S. INSTITUTIONS

There are at least ten international programs in the United States dealing with global management issues (Table C.4). Three of these (Columbia, Harvard, and Penn State) are in the general management category and have been in place for many years. The other seven are relatively new offerings in the international areas.

Flexibility and Special Features of U.S. General Management Programs

U.S. general management programs are changing in design and content by the introduction of options and special features to meet the varying needs of executives who attend these courses. While many of the following features have been in place in most programs for some time, a significant number of leading business schools are introducing these changes and additions in their 1991 offerings for the first time. Furthermore, there is little doubt that other executive programs will follow suit in the years to come.

Some special features in U.S. general management programs are computer simulations, spouse programs, and physical fitness sessions. The majority of the sixty-three leading general management programs already include these three features. However, the trend indicates an increasing focus on spouse attendance and cardiovascular fitness options. The principal special features are optional sessions and electives; Outward Bound experiences; leadership assessment sessions; special projects; and CEO and other executive speakers.

OPTIONAL SESSIONS AND ELECTIVES (10 PROGRAMS)

1. Columbia University: Executive Program in Business Administration/Managing the Enterprise
2. Columbia University: Executive Program in International Management
3. Harvard Business School: Program for Management Development
4. University of Minnesota: Minnesota Executive Program

5. University of Minnesota: Minnesota Management Institute
6. Massachusetts Institute of Technology: Sloan Fellows Program
7. University of Pittsburgh: Management Program for Executives
8. Stanford University: Advanced Management College
9. Stanford University Stanford Sloan Program
10. Wharton School, University of Pennsylvania "Advanced Management Program"

OUTWARD BOUND EXPERIENCES (8 PROGRAMS)

1. Boston University: Leadership Institute
2. Boston University: Management Development Program
3. Carnegie-Mellon University: Program for Executives
4. Dartmouth College: Tuck Executive Program
5. University of Houston: Executive Development Program
6. University of Minnesota: Minnesota Executive Program
7. University of Minnesota: Minnesota Management Institute
8. The Ohio State University: Executive Development Program

LEADERSHIP ASSESSMENT SESSIONS (8 PROGRAMS)

1. University of Arizona: Executive Development Course
2. Boston University: Leadership Institute
3. Carnegie-Mellon University: Program for Executives
4. Columbia University: Executive Program in Business Administration/Managing the Enterprise
5. University of Houston: Executive Development Program
6. Massachusetts Institute of Technology: Program for Senior Executives
7. University of Pittsburgh: Management Program for Executives
8. Wharton School, The University of Pennsylvania: Advanced Management Program

SPECIAL PROJECTS (7 PROGRAMS)

1. Columbia University: Executive Program in Business Administration/Managing the Enterprise
2. University of North Carolina: The Executive Program
3. Northeastern University: Executive Development Program
4. Northwestern University: Advanced Executive Program
5. Smith College: Smith Management Program

6. University of Southern California: International Business Education and Research Program
7. Wharton School, The University of Pennsylvania: Advanced Management Program

CEO AND OTHER EXECUTIVE SPEAKERS (11 PROGRAMS)

1. University of California-Berkeley: The Executive Program
2. Carnegie-Mellon University: Program for Executives
3. Harvard Business School: International Senior Management Program
4. Massachusetts Institute of Technology: Sloan Fellows Program
5. The Ohio State University: Executive Development Program
6. University of Pittsburgh: Management Program for Executives
7. Simmons College: Program for Developing Executives
8. Smith College: Smith Management Program
9. Stanford University: Stanford Executive Program
10. University of Southern California: International Business Education and Research Program
11. Wharton School, The University of Pennsylvania: Advanced Management Program

University-Corporate Relationships

Joint Research Activities

Business schools and their corporate clients are developing much closer relationships in terms of joint research efforts to further understand the future requirements of effective executive education. Some recent examples of such projects are:

The International Consortium for Executive Development Research. This new effort brings together members from the world's leading business schools and major corporations to explore new approaches to global leadership and to determine the best practices in executive development around the world.

Boston University's Executive Development Roundtable. This consortium brings together seasoned human resources development professionals from major world corporations to explore cutting-edge approaches to the development of executives throughout their careers.

Boston University's Manufacturing Executive Forum. The Forum brings together multifunctional executives from leading manufacturing organizations to address the issues of manufacturing competitiveness and quality.

Carnegie-Mellon's Carnegie-Bosch Institute of Applied International Management. Having a dual mission, this organization conducts research on international management issues and develops new executive education programs.

Duke Fuqua School of Business. This school conducts field research in Europe, Latin America, and, soon, in the Far East on behalf of worldwide client companies.

Special Focus Programs

In addition to the growing number of corporate in-house executive programs conducted by universities on a subcontract basis, many leading business schools ar mounting a range of industry-specific and special focus programs to meet a particular set of needs. Examples include:

Carnegie-Mellon's special program for research and development managers in the Bell Laboratories to cover business operations and how R&D can contribute to competitive advantage. Although started in 1984, this program has been run six times each year and is the most highly rated in Bell's custom program catalogue.

Duke Fuqua School of Business's offerings oriented to specific needs of the telecommunications industry, the legal profession, and accounting firms, in addition to a number of tailored, company-specific programs. Another new Duke program is one for Soviet managers. They are enrolled in a four-week general management session with three weeks of classes and a one-week internship. Some 129 Soviet managers participate each year.

University of Wisconsin-Madison's special program for middle managers in two of the largest local employers: the CUNA Mutual Insurance Group and the Wisconsin Physicians' Service. This unique program has been extremely successful in meeting the needs of middle managers in those two organizations.

Massachusetts Institute of Technology's Protocol Agreement with two major institutions in the U.S.S.R. to cooperate in areas of executive education. MIT also offers a special program for corporate legal officers dealing with the strategic legal concerns facing large companies.

University of Virginia's Darden School's industry-specific offerings for the aviation and banking sectors, in addition to its many tailored company programs.

Length of Executive Programs in the United States

One of the significant trends in executive education in recent years has been the shortening of programs, particularly in the general management category. This action is primarily due to changing business conditions, a reduction of training budgets, and corporate executive needs to minimize time away from the job.

Programs shortened since 1980 include Columbia University's Program in Business Administration, from six to four weeks; Emory University's Advanced Management Program, from six to four weeks; Harvard Business School's Advanced Management Program, from thirteen to eleven weeks and its Program for Management Development from fourteen to twelve weeks; University of Pittsburgh's Management Program for Executives, from six to five weeks; and Williams College's Williams Executive Program, from six to five weeks.

Programs to be shortened in 1992 include Carnegie-Mellon University's Program for Executives, from six to four weeks; Cornell University's Executive Development Program, from five to four weeks; University of Pittsburgh's Management Program from five to four weeks; and Stanford University's Stanford Executive Program, from eight to seven weeks.

TABLE C.1 Overseas Executive Programs by U.S. and European Institutions

	Program	Location	Duration	Participant Level and Mix
The University of Michigan	Global Leadership Program (1989)	U.S., Brazil, India, China	5 weeks	Middle managers from the U.S., Japan, Brazil, Britain, and India
The University of Pennsylvania	International Forum (1989)	Tokyo, Japan, Paris, France, and Philadelphia, PA	Three 1-week units	General managers from national and international organizations
Stanford University/ National University of Singapore	General Management Program (1983)	Singapore, Malaysia	3 weeks	Senior managers from international organizations
The University of Virginia	1. Management Program (1979)	Little Bay, Australia	2 weeks	Upper-middle and senior managers from Australia
	2. Managing Critical Resources (1988)	Cambridge, England	2 weeks	Upper-middle and senior managers from England
INSEAD	1. Asian International Executive Programme (1981)	Singapore, Malaysia	2 weeks	Upper-middle managers from international organizations
	2. Hong Kong International Executive Programme (1981)	Macao	3 weeks	Upper-middle managers from international organizations
	3. Strategic Planning in Asia (1986)	Singapore, Malaysia	1 week	Upper-middle managers from international organizations

TABLE C.1 *Continued*

Program	Location	Duration	Participant Level and Mix
4. Asian International Marketing Programme (1987)	Macao	2 weeks	Middle managers from international organizations
5. Human Resources Management in Asia (1984)	Macao and Singapore, Malaysia	1 week	Asian and non-Asian middle managers
London Business School Competing Globally: The View from Japan	Tokyo, Japan	2 weeks	Senior executives from international organizations
Aspen Institute for Humanistic Studies Seminars and conferences	Berlin, France, Italy, Stockholm	Several days	Senior executives and leaders from various countries
Boston University Managing Global New Product Development (1989)	European cities	10 days	Senior and upper-middle managers from the U.S. and overseas
The Brookings Institution 1. Europe—1992 (1985)	Paris, France	1 week	Primarily senior executives from the U.S.
2. A Look Inside Policy-Making in Japan (1983)	Tokyo, Japan	1 week	Senior U.S. executives with international responsibilities

TABLE C.1 *Continued*

	Program	Location	Duration	Participant Level and Mix
The University of California-Berkeley	1. Advanced Management Program for Caribbean and South America (1985)	Kingston, Jamaica	2 weeks	Upper-middle and senior managers from the U.S. and Latin America
	2. Advanced Management Program for Asia-Pacific Managers (1986)	Bangkok, Thailand	2 weeks	Upper-middle and senior managers from the Asia-Pacific area
	3. International Strategic Management for Asia-Pacific Executives (1989)	Singapore, Malaysia	1 week	Upper-middle and senior managers from the Asia-Pacific area
	4. Asia-Pacific Corporate Financial Management Program (1987)	Singapore, Malaysia	1 week	Upper-middle and senior financial managers from the Asia-Pacific area
	5. Competitive Marketing Strategies for Asia-Pacific Managers (1989)	Singapore, Malaysia	1 week	Upper-middle and senior marketing managers from the Asia-Pacific area
	6. Managing Technology-Based Companies and Products for Asia-Pacific (1989)	Singapore, Malaysia	1 week	Upper-middle and senior managers from the Asia-Pacific area

TABLE C.2   International Participation in U.S. General Management Programs

	Program	Duration	Participant Level and Mix	International Attendance (%)
Harvard Business School	International Senior Management Program (1973)	9 weeks	Senior managers from large multinational corporations	90
Columbia University	Executive Program in Business Administration/Managing the Enterprise (1952)	4 weeks	Senior managers from national and international companies	60
Columbia University	Executive Program in International Management (1960)	4 weeks	Upper-middle management from 35 countries	60
University of California-Berkeley	The Executive Program (1959)	4 weeks	Upper-middle management from diverse organizations	50
Massachusetts Institute of Technology	Program for Senior Executives (1956)	9 weeks	Senior managers from the U.S. and overseas	50
Stanford University	Stanford Executive Program (1952)	8 weeks	Senior managers from the U.S. and overseas	45
Wharton School, The University of Pennsylvania	Advanced Management Program (1988)	5 weeks	Senior management from a broad range of organizations	45
Harvard Business School	Advanced Management Program (1943)	11 weeks	Senior executives from large corporations	40
University of Pittsburgh	Management Program for Executives (1949)	5 weeks	Upper-middle managers from U.S. and international firms	40

TABLE C.2 *Continued*

	Program	Duration	Participant Level and Mix	International Attendance (%)
University of Houston	Executive Development Program (1953)	4 weeks	Upper-middle managers from the U.S. southwest and overseas	30
University of Illinois	Executive Development Program (1957)	4 weeks	Upper-middle managers from a range of U.S. and overseas firms	30
Northwestern University	Advanced Executive Program (1951)	4 weeks	Senior executives from large U.S. and international firms	30
The Pennsylvania State University	Executive Management Program (1956)	4 weeks	Mostly upper-middle managers from a broad range of firms	30
Carnegie-Mellon University	Program for Executives (1954)	6 weeks	Upper-middle management from large and mid-sized firms	25
University of Michigan	The Executive Program (1954)	4 weeks	Senior and upper middle management from the U.S. and abroad	25
University of Virginia	The Executive Program (1958)	6 weeks	Senior management from the U.S. and overseas	25
Cornell University	Executive Development Program (1953)	5 weeks	Upper-middle management from the U.S. and overseas	20

TABLE C.3 International Participation in U.S. Functional and Specialized Programs

	Program	Duration	Participant Level and Mix	International Attendance (%)
Finance				
Columbia University	Financial Management Program (1982)	6 days	Financial and line officers at senior and upper-middle levels	30
Massachusetts Institute of Technology	Executive Program in Financial Management (1984)	1 week	Senior financial executives from large organizations	30
Harvard Business School	Corporate Financil Management (1973)	2 weeks	Senior managers from both the financial and nonfinancial areas	25
Stanford University	Financial Management Program (1976)	2 weeks	Senior financial staffs and line managers	20
Marketing				
The Pennsylvania State University	Industrial Marketing Management Program (1976)	2 weeks	Upper-level executives with product marketing responsibility	40
Harvard Business School	Strategic Marketing Management (1972)	2 weeks	Senior managers with marketing responsibilities from large companies	30
The Pennsylvania State University	Industrial Sales Management Program (1976)	1 week	Upper-middle managers from industrial sales backgrounds	20
Stanford University	Marketing Management Program (1976)	2 weeks	Experienced senior marketing executives from large companies	20
Operations				
Harvard Business School	Manufacturing in Corporate Strategy (1975)	2 weeks	Senior executives with multiple plant or corporate responsibilities for operations	30

187

TABLE C.3 *Continued*

	Program	Duration	Participant Level and Mix	International Attendance (%)
(Operations, continued)				
The Pennsylvania State University	Manufacturing Strategy and Technology Program (1978)	2 weeks	Mostly upper-middle managers from large corporations	30
The Pennsylvania State University	Program for Logistics Executives (1980)	10 days	Senior and upper-middle managers from logistics functions areas	25
Technical				
Massachusetts Institute of Technology	Management of Technology Program (1981)	1 year	High potential middle managers from technical areas	55
Massachusetts Institute of Technology	Management of Research, Development, and Technology-based Innovation (1964)	2 weeks	Upper-middle managers from research and engineering	50
Massachusetts Institute of Technology	Current Issues in Managing Information Technology (1975)	3½ days	Upper-middle management from information systems areas	25
Harvard Business School	Managing the Information Services Resource (1972)	2 weeks	Senior executives from the information systems areas	20
University of California- Los Angeles	Managing the Information Resource (1980)	1 week	Upper-middle managers from information systems areas	20
	Specialized			
Human Resources Development				
The Pennsylvania State University	Human Resources Management Program (1979)	2 weeks	Human resource professionals at the upper-middle management level	45
Stanford University	Executive Program in Organizational Change (1981)	2 weeks	Senior line management with some human resources staff	20
Organizational Behavior				
The Pennsylvania State University	Program for Strategic Leadership (1977)	2 weeks	Upper-middle managers from a variety of organizations	40

TABLE C.3 *Continued*

	Program	Duration	Participant Level and Mix	International Attendance (%)
(Organizational Behavior, continued)				
Southern Methodist University	Management of Managers (1984)	2 weeks	Upper-middle and senior managers from a range of line and staff	30
University of California-Berkeley	Management Development (1977)	1 week	Middle managers from operations and human resource functions	20
University of Chicago	Management Development Seminar (1957)	3 weeks	Senior and upper-middle managers from a broad range of functions	20
Strategy				
Columbia University	Business Strategy Program (1974)	2 weeks	Upper-middle managers who manage strategic planning processes in large organizations	40
The Pennsylvania State University	Strategic Purchasing Management Program (1963)	10 days	Upper-middle and senior managers with strategic purchasing responsibilities from mostly large corporations	40
Massachusetts Institute of Technology	Executive Program in Corporate Strategy (1972)	1 week	Experienced senior executives from a broad range of national and international companies	35
University of California-Los Angeles	Advanced Executive Program (1986)	2 weeks	Senior executives with international strategy responsibilities	30
Wharton School, The University of Pennsylvania	Strategic Management (1984)	5 days	Mostly upper-middle management from a wide range of organizations	20

189

TABLE C.4 International Programs Offered at U.S. Institutions

	Program	Location	Duration	Participant Level and Mix
Columbia University	Executive Program in International Management: Managing for Global Success (1960)	Harriman, NY	4 weeks	General managers from multinational companies from all over the world
Harvard University	International Senior Management Program (1973)	Boston, MA	9 weeks	Senior executives who work in an international environment
The Pennsylvania State University	Managing the Global Enterprise Program (1986)	University Park, PA	2 weeks	Executives from all nations who have international responsibilities
Babson College	Achieving Global Integration (1990)	Wellesley, MA	1 week	Senior managers who have strategic responsibilities
Simmons College	International Management for Women (1990)	Boston, MA	6 weeks	Experienced women managers
Massachusetts Institute of Technology	Executive Program in Japanese Technology Management (1990)	Dedham, MA	1 week	Senior line and staff managers from technical and nontechnical fields
University of Michigan	Strategies for Global Competition (1989)	Ann Arbor, MI	1 week	Managers and executives who have global competitors and those who have foreign operations
University of Illinois	Specialized Program for International Managers (1990)	Urbana, IL	4 months	Upper-middle managers who need an understanding of international management
George Washington University	Managing in the Global Marketplace (1989)	Washington, DC	1 week	Upper-middle managers who need understanding of interbusiness issues
University of Washington	International Management: Doing Business in Japan (1989)	Seattle, WA	1 week	American middle managers who need understanding of Japan

APPENDIX D

History of Executive Education in America

The Early Years

In the late nineteenth century the management of business enterprise was becoming recognized as a crucial factor in the national welfare and a demanding function for its practitioners. The concept of management as a distinct kind of work and field of learning was emerging, and with it a new managerial class within the social structure. As a consequence some universities and business leaders saw the need for special kinds of education that were not provided by the classical liberal arts and natural sciences programs.

In 1881 the University of Pennsylvania established the Wharton School of Commerce and Finance. The universities of Chicago and California followed suit before 1900. Dartmouth established the Amos Tuck Graduate School of Business in 1900. In 1908 Harvard required a bachelors degree in administration for admission to its two year course leading to an MBA. During 1911 to 1912 Dartmouth sponsored a series of seminars for businessmen based on Frederick Taylor's work in "scientific management." This series may have been the first widely known educational sessions solely for practicing business executives.

America emerged from World War I as a creditor nation, and business boomed. Undergraduate business schools proliferated and more graduate schools appeared. But for years there was no sustained move toward executive education programs for mid-career business people. However, theories and experiments in industrial psychology and organization were evolving. They would soon be put to use in the classroom and further developed in practice.

In 1928 the Harvard Business School started conducting summer sessions for experienced businessmen. In 1931 the year-long Sloan

Fellowship Program was started at M.I.T., which, along with its sister program at Stanford, still continues. In 1935 Harvard initiated a series of executive discussion groups, the so-called Philip Cabot Weekends. In 1941 an executive education program was established at the University of Iowa. In 1943 the federal government asked Harvard and Stanford to retrain business executives in the intricacies of the war production effort. At Harvard the program was gradually transformed into a general course of study, and in 1945 it was renamed the Advanced Management Program and continues to this day.

The Postwar Years

The 1930s Great Depression followed by World War II had created a pent-up demand for high quality entrepreneurial talent at all managerial levels. New technologies, markets, and financial innovations were changing the face of business itself.

General management programs were started at Dartmouth, Western Ontario, Pittsburgh, Toronto, Indiana, and Stanford universities. The American Management Association, founded in 1923, introduced computer-program management exercises into its executive programs. Harvard and Western Ontario developed the use of small group studies and business cases, techniques borrowed from law schools. The Aspen Institute for Humanistic Studies focused in its programs on the problems of power and responsibility, topics that are now treated in many other executive programs. Large companies were establishing their own internal programs, usually to supplement those offered publicly.

A study by the U.S. Naval Institute in 1948 reported that executive education was an active component in the executive development efforts of more than 100 medium- to large-size companies. By 1958 there were forty residential university programs two weeks or more in length, conducted frequently each year. Between 1949 and 1958 there had been roughly 10,000 attendees from over 2,700 companies.

The Maturing of Executive Education

Almost from its beginnings, executive education in America had derived most of its intellectual thrust from the established faculties of colleges and universities. Teaching faculties, not only in business schools but also in more traditional disciplines, found this new market to be professionally stimulating and lucrative.

By 1958 there were seven exclusively graduate schools of business: Chicago, Columbia, Cornell, Dartmouth, Harvard, Pittsburgh, and Stanford. In addition, more than 100 universities offered some graduate courses in business. There were hundreds of shorter courses, seminars, workshops, and specialized functional programs. For example, Michigan State carried 130 offerings, and there were eight nonresidential broad-coverage programs taking a year or longer. At least ten liberal arts courses had been devised for executives by 1958, but with only Wabash, University of Denver, and the Aspen Institute surviving at that time. Later surveys reported that, except for several temporary downturns, annual attendance at the longer general management programs had risen from about 3,900 in 1969 to 5,800 to 1979.

A National Industrial Conference Board report in 1969 listed forty-five programs of two to sixteen weeks in length. It noted that 66 percent of those programs were two to four weeks long. Most were continuous, but some were split session offerings. Ninety percent of the university brochures listed two common objectives: to make generalists out of specialists, and to increase the attendees' effectiveness through their exposure to decision making, communications, and behavioral science findings. Seventy-one percent said that they tried to acquaint their participants with forces in the external environment of business. Thirty-three percent included methods of research and statistical analysis in their curricula. Other program objectives were to broaden and deepen business understanding, provide opportunities to discuss ideas with people in other organizations, and to allow participants to reflect on their career development and life planning.

Recent Trends

Certain trends in executive education have appeared over the past several decades (see also Appendix C). For instance, executive education has become more institutionalized, an integral part of the university structure in some cases. Likewise, companies have become increasingly committed to its support, with some of the larger ones investing in their own campuses and programming staffs.

Course content and teaching methods have become more pragmatic. Teaching resources and course materials from business and other institutions are increasingly being incorporated into program structure. Behavioral sciences content has became more practice-oriented and less theoretical. Environmental and global aspects of management are receiving increased attention. The teaching of strategy, planning and control, and the financial aspects of business has become more sophisticated.

The attendee groups show increasing numbers of women, ethnic participants, and people from other nations. A wider selection of programs has become available as sponsoring organizations seek to appeal to special interests and different age and experience groups. Also it seems that, on the whole, participants are taking the study programs more seriously than in times past, when businesses and their middle-manager populations were under comparatively less national and international competitive pressure.

We can't help but note that a certain correspondence exists between the such trends and what seems to be happening in the nation and throughout the world. For example, U.S. executive education programs are beginning to reflect the lessons that we are slowly and painfully learning from the managerial and technological advances in other countries. It has been said that the institution of education reflects more than molds the interests, priorities, and values of the society which it serves. The newest and most rapidly changing segment of that institution, executive education, seems to confirm the truth of that statement.

REFERENCES

Andrews, K. 1966. *The Effectiveness of University Management Programs.* Harvard University, Cambridge, MA.

Barney, G. 1980. *The Global 2000 Report to the President of the U.S. Council on Environmental Quality and Department of State.* Pergamon Library, Washington, DC.

Berman, J. 1981. *Study of Participants of Leading University Management Development Programs.* ABT Associates, Cambridge, MA., and The Union Carbide Corporation, Danbury, CT.

Billy, C. 1991. *The Bricker Bulletin.* Peterson's Guides, Princeton, NJ.

Byrne, J. 1991. Effective Education Survey. *Business Week* (Oct. 28):102–107.

Crotty, P. 1974. *Continuing Education and the Experienced Manager.* Bureau of Business and Economic Research, Northeastern University, Boston, MA.

Deal, T., and A. Kennedy. 1982. *Corporate Cultures.* Addison-Wesley, Reading, MA.

DuJardin, P. 1985. The Effects of Advanced Management Programs, Parts One and Two. *Management Review* (Aug.):44–56.

Fresina, A. 1988. *Executive Education in Corporate America.* Executive Knowledgeworks, Palatin, IL.

Gibson, D., and D. Jones. 1979. *Survey of Management Development Programs.* Harbridge House, Boston, MA.

Ingolls, C. 1986. Executive Education Programs: Meeting Strategic Organizational Purposes. Ph.D. dissertation, Harvard University, Cambridge, MA.

Jenner, S. 1982. *Attitude Change during Management Development Programs Based on a Longitudinal Study of the Advanced Management Program at the University of Hawaii, 1978–1982.* The University of Hawaii, Honolulu, HA.

Jennings, E. 1980. *Mobility Management.* Special Report, Michigan State University, East Lansing, MI.

Lerner, M. 1960. Introduction, in *The Prince* by Niccolo Machiavelli (1513), translated by W. K. Marriott. Dutton, New York.

McCall, M., and M. Lombardo, 1988. *The Lessons of Experience.* The Center for Creative Leadership, Lexington Books, D. C. Heath and Company, Lexington, MA.

McGregor, D. 1960. *The Human Side of Enterprise.* McGraw-Hill, New York.

Newland, C. 1976. *Evaluation of the Federal Executive Institute: A Report of Executive Perceptions*. Federal Executive Institute, Charlottesville, VA.

Roche, G. 1979. Much Ado about Mentors. *Harvard Business Review* (Jan.-Feb.).

Saari, L. 1987. The GMAC/Battelle Study of University-based Executive Education. Presentation at The Consortium for University Executive Development Programs in November, 1988, New York.

Schein, E. 1967. Attitude Change during Management Education. *Administrative Science Quarterly,* 601–628.

Schrader, A. 1985. How Companies use University-based Executive Development Programs. *Business Horizons* (March-April):53–62.

Silk, L. 1980. Workforce 2000. *New York Times,* Aug. 25.

Sparks, D. 1979. Evaluating an Existing Executive Development Program in a College of Business Administration, in *Determining the Payoff in Management Training,* pp. 83–112. American Society for Training and Development, Madison, WI.

Vicere, A. 1983. *Executive Education in the Fortune 500 Companies, 1981–1982 Survey Reports*. Office of Executive Programs, The Pennsylvania State University, University Park, PA.

Vicere, A., and V. Freeman. 1990. Executive Education in Major Corporations: An International Survey. *Journal of Management Development.*

Viteles, M. 1956. An Evaluation of the Bell Program at the University of Pennsylvania, in *Towards the Liberally Educated Executive*. The Fund for Adult Education, White Plains, NY.

Wack, P. 1985. *Scenarios: The Gentle Art of Re-Perceiving*. Division of Research, Harvard Business School, Cambridge, MA.

Watson, R. 1973. Addressing the Effectiveness of a University Executive Management Program: A Field Experiment. Ph.D. dissertation, University of Illinois, Urbana, IL.

ADDITIONAL READINGS

1974. *Survey of External Management Education Programs*. AT&T Human Resources Development Staff, New York.

1979. *Industry 2000: New Perspectives*. United Nations Industrial Development Organization, Vienna, Austria.

1980. Shell's Scenario Planning. *World Business Weekly* (April 7).

1985. *Over-all Socio-Economic Perspective of the World Economy to the Year 2000*. Report of the Secretary General, United Nations General Assembly, New York.

1985. What-if Shadows in the Crystal Ball. *The Economist* (July 20).

1985. The Year 2000. *Money Magazine* (Oct.).

1986. University-based Executive Education Programs: A Peer Evaluation. *Personnel Administrator* (Feb.):117–118.

1987. Demographic Forces. *Providence-Journal Bulletin* (Oct. 4).

1987. Older America. *Modern Maturity* (Oct.-Nov.).

1987. Remaking the American CEO. *New York Times* (Jan. 25).

1988. *Management for the Future*. Management Research Group, Ashridge Management College, Berkhamsted, UK.

Blank, M. 1986. *Management Training and Executive Education in Large Corporations*. Bell Atlantic Report, Silver Spring, MD.

Cox, A. 1982. *The Cox Report on the American Corporation*. Delacorte Press, New York.

Grayson, L., and A. Spencer. 1981. *Executive Profile of the Top Officers of the Largest Oil Companies in the U.S., 1950–1980*. Energy Policy Study Center, University of Virginia, Charlottesville, VA.

Peters, T., and N. Austin. 1985. *A Passion for Excellence*. Harper & Row, New York.

Peters, T., and R. Waterman. 1982. *In Search of Excellence*. Harper & Row, New York.

Porter, L., and L. McKibbin. 1987. *Management Education in the 21st Century*. American Academy of Collegiate Schools of Business, Washington, DC.

Rada, J. 1986. *Report of the Commission for the Year 2000*. International Management Institute, Geneva, Switzerland.

Schwartz, P. 1984. *Scenario, The Challenge of the Future—Alternatives to Pessimism*. Shell World, London, UK.

Schwartz, P. 1985. *Scenarios, 1984–2005*. Business Environment and Group
    Planning, Royal Dutch Shell Group, London, UK.
Toffler, A. 1980. *The Third Wave*. William Morrow & Co., New York.
Verlander, E. 1985. Use of Principles of Adult Learning in University Executive
    Programs. Ph.D. dissertation, Columbia University, New York.
Verlander, 1988. Executive Transformation Programs. *Training and Develop-
    ment Journal*.
Verlander, E. 1990. The Executive Learner. *Journal of Management Develop-
    ment* **9**(4):4–6.
Vicere, A. 1987. *Executive Education and Development in the 21st Century*.
    The Pennsylvania State University, University Park, PA.
Vicere, A. 1988. *University-based Executive Education: Impacts and Implica-
    tions*. The Pennsylvania State University, University Park, PA.
Vicere, A. 1989. *Executive Education: Process, Practice and Evaluation*. Pe-
    terson's Guides, Princeton, NJ.

INDEX

Age and obsolescence: aging process, 39; obsolescence, 45–47; older managers, 46–47

Assessment: assessment centers, 44, 130; executive program participant assessment, 75, 125–130; self-assessment, 75

Assignments: assistant-to, 22–23; and career planning, 30–32; developmental line, 25; international, 23–25; larger scope, 26; line-to-staff, 21–22; project and task force, 20–21; start-up, 25; turnaround, 25–26. *See also* On-the-job development

Attitude: corporate, 75, 93–94; executive, 93; student, 72–74, 78–79. *See also* Student

Business: activity, 6, 156–164; foreign ownership of, 160; globalization of, 160–162; ownership of, 6; scope of, 12; structure of, 156–158

Career problems: criteria for success, 35–38; derailing, 41–45; managing disappointment, 38, 63; mentors, 33–34; and midlife crises, 39; plateauing, 39–40; success versus failure, 51–53. *See also* Perspective

Careers: career manager, 24–25; dummy theorem, 43; functional, 32–33; multiple-company, 31; out-placement, 41; planning, 30–32; promotion from within, 30; single company, 31; stages of, 59–63

Change: consequences of, 57–58; historical, 57, 142; in managerial perspective, 59; in modes of thinking, 88–91; process and examples of, 4–7

Competencies: executive, 11; leadership, 141; managerial level, 16; mobility and, 32–33; requirements, 14;

scenarios and, 14; trends supporting, 12–13. *See also* Program content

Corporate: attitude, 75, 93–94; culture, 97, 132; make-or-buy policy, 94–95; motivation for internal programs, 96–98; needs, 95–96; staff, 29

Executive education: characteristics of, 55–56; evolution of, 54–55; flexibility and special features of, 177–179; globalization of, 176–177; growth of in the U.S., 175–76; history of, 191–194; a liberal, 84; limitations of, 64–65; research, 169–174; and university/corporate relations, 179–180

Executive programs, categories of: degree, 110; external, 106–110; functional, 106–108; general management, 106; internal, 96–98; special focus, 110; specialized, 108–109

Executive programs, external: credentials, 111; definition of, 105–106; evaluation of, 124–125; formal, 53; joint responsibility for, 131–134; return on investment, 124; selection process, 117–120. *See also* Executive programs, categories of

Executive programs, internal: cautions about, 101; design, integration, and administration, 101–103; faculty resources, 98–99; interface with the chief executive, 100; managers or directors of, 100–101; motivation for, 94, 96

Executives: attitudes, 14, 72–74; the twenty-first century, 11–12, 14. *See also* Executive education

Forecasting: risks, 3–4, 147–148. *See also* Scenarios

Future: trends, 148–167, 175–181

199